communication skills

communication skills

What the best communicators know, do and say

Gill Hasson

Harlow, England • London • New York • Boston • San Francisco • Toronto • Sydney • Auckland • Singapore • Hong Kong
Tokyo • Seoul • Taipei • New Delhi • Cape Town • São Paulo • Mexico City • Madrid • Amsterdam • Munich • Paris • Milan

PEARSON EDUCATION LIMITED

Edinburgh Gate
Harlow CM20 2JE
Tel: +44 (0)1279 623623
Fax: +44 (0)1279 431059
Website: www.pearsoned.com/uk

First published in Great Britain in 2012

ISBN: 978-0-273-76174-7

British Library Cataloguing-in-Publication Data
A catalogue record for this book is available from the British Library

Library of Congress Cataloging-in-Publication Data
A catalog record for this book is available from the Library of Congress

10 9 8 7 6 5 4 3 2 1
15 14 13 12 11

Typeset in 10/14pt Plantin by 3
Printed and bound in Great Britain by Henry Ling LH.,
at the Dorset Press, Dorchester, Dorset

Contents

About the author

Gill Hasson is an associate tutor at the University of Sussex where she teaches a course on personal and career development. She is also a freelance trainer and delivers courses for voluntary organisations, schools, parents, young people, social workers and youth workers. These courses include assertiveness and confidence building, communication skills and communicating with children and teenagers. She has written courses on the subjects of health and child development for the Open University.

Gill is a co-author of *Bounce: Use the power of resilience to live the life you want* (Pearson, 2009) and *How to be Assertive* (Pearson, 2010). She also writes for magazines such as *Psychologies* and *Natural Health*.

Gill can be contacted at gillhasson@btinternet.com

Acknowledgements

Thanks to Roz, Lennie, Phil, Gilly, Wendy, Sarah, Donna, Janine, Harry, Tom and Andy.

Introduction

Every day, you interact with people who have different opinions, beliefs, feelings and needs from you. Your ability to exchange ideas and opinions with other people, understand their point of view and solve problems between you will depend on how effectively you are able to communicate.

Effective communication is important in all areas of your life: at work, at home, with friends and family and with total strangers. Good communication skills make you interesting! They can help you connect with people, build friendships and develop relationships at work and at home.

Being able to communicate more effectively with *family and friends* means that you will know what to say to get others to open up to you; you will be able to understand their feelings and emotions, what they mean and what they need. In turn, you are more likely to be listened to and understood.

In a range of situations, effective communication enables you to build a rapport easily with new people and feel confident about communicating with a wide range of people.

Being able to communicate more effectively at *work and in professional settings* gets you ahead. If you are an effective communicator, you will appear more organised and knowledgeable and increase your chances of getting where you want to in your work and career. An effective communicator can easily explain useful and interesting ideas and meet other people's responses – whatever they are.

Others will take you seriously. They will listen to you and engage in discussion.

At work, ineffective communication serves as an obstacle to getting things done; ineffective talking leads to ineffective action! But good communication skills can minimise misunderstandings and confusion: they can save time by enabling you to get relevant information easily and quickly.

Finally, having good communication skills will increase confidence and trust between colleagues at all levels, and you will enjoy better working relationships.

Convinced? Now you just need to know how to get those all-important skills.

This book is divided into two parts. In Part 1 you will learn what those skills are and how to develop them; skills such as active listening, assertive speaking and clear, concise writing. Part 2 discusses specific situations and shows how those communication skills work in practice.

Chapter 1 starts by describing what communication is and what makes for brilliant communication skills. Knowing how to improve your communication skills will be easier once you are aware of your own communication style; how you express yourself and how others understand you. The quizzes in this chapter will help you to identify your communication approaches, attitudes and strengths. You'll also learn how and why a range of differences between you and other people can create barriers to communication.

But whether you're trying to tell someone you don't understand what they're talking about or strike up a conversation at the bus stop, there are two qualities that you need: *confidence* and *empathy*. There are ideas and tips in Chapter 2 to help you develop both.

You will discover that no matter what levels of empathy and confidence you currently have, the best place to develop your communication skills is from a position of strength.

However, even the most confident communicator can have their communication skills undermined. When, for example, expectations and emotions get in the way, strengths, confidence and empathy can disappear and you can find yourself falling back on inappropriate and unhelpful behaviour.

This chapter emphasises the importance of using your emotional intelligence; letting your emotions *inform* what you say, *not* take over what you say!

In Chapter 3 the focus is on active listening. Active listening is possibly the most important communication skill that you could learn; it's the quickest route to developing rapport, empathy and understanding. You will learn how to be an active listener in this chapter and you will see that active listening has many benefits including, for example, helping you to ask the right questions; to encourage the other person to open up and say more.

Although active listening is a crucial skill in developing brilliant communication skills, the ability to 'read between the lines' will give you a further advantage.

Certainly, as you will see, claims that communication is 93% non-verbal and only 7% verbal are inaccurate. But, what *is* beyond question is that non-verbal behaviour is a powerful communicator.

In Chapter 4 you will learn some of the many ways in which we all communicate non-verbally, so that you can use and understand these signs and signals to communicate more effectively. You will discover that the meanings of non-verbal communication differ from one person to the next, and there are many cultural differences, too. This chapter also stresses the fact that it's not always possible to interpret a single gesture or signal in isolation; you need to look for 'clusters'.

However and whatever you want to say, there's one key principle: you have to communicate in a way that makes it easy for other people to listen and to understand what you mean. Chapter 5 explains how to communicate purposefully; what to say, how to say it and when to say it.

Of course, when you communicate in person or on the phone, if there are any misunderstandings you can clarify what you mean immediately. When you communicate in *writing*, this doesn't happen, so what you say in writing must express *exactly* what you mean – there is nothing else to support or clarify your intentions, feelings, requests, information and ideas.

At work and in business, well-written letters, emails, reports and other communications help to create a good impression and can prove valuable in increasing business. How do you communicate effectively in writing? The same as you do when speaking: clearly and concisely, with confidence and empathy. Chapter 6 explains how.

Throughout Part 1, tips and techniques, practical pointers and examples are included to help you communicate easily and effectively in any situation. However, communication is rarely straightforward. Often, it's dependent on context; the circumstances and situation in which communication is happening. Part 2 looks at some specific situations, such as how and what to say when you are giving advice or making small talk. It highlights the importance of good non-verbal communication during an interview or when giving a presentation. There's also advice to ensure that your emails are clear, concise and polite – even if your response is angry!

In fact, when things are going badly, communication can be as hard as wading through wet concrete. Part 2 shows you how to communicate in difficult situations. Scenarios that include handling criticism and bad news are potentially difficult for everyone concerned. But they *can* be handled with confidence and sensitivity; the last scenario in the book is a good example: how to get someone to shut up. Nicely!

And that is pretty much the theme for the whole book – not just brilliant communication, but *positive* communication.

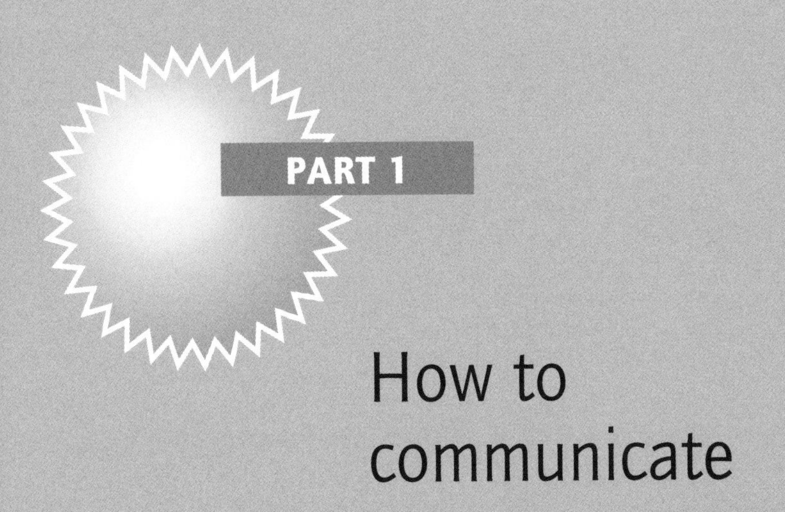

PART 1

How to communicate

What is communication? Knowing your communication style

What communication is and what effective communication skills are

What exactly is 'communication'? Depending on whom you ask, the answer may vary. Communication is:

- discussion of ideas between one person and another;
- talking to or socialising with other people;
- interaction between people;
- getting your ideas across;
- talking and responding to someone.

Of course, it's all of these things (and, interestingly, the answer each person gives provides a clue to their communication style – more about that later in this chapter).

The term 'communication' comes from the Latin word 'communis' which means 'common'. So, 'to communicate' means 'to make common' or 'to make known'. This act of making common and known is carried out through the exchange of thoughts, ideas and feelings.

> communication is the *exchange* of ideas, information and feelings

Communication, then, is the *exchange* of ideas, information and feelings. Note the emphasis is on the word 'exchange'; exchange involves giving *and* receiving. So communication is a two-way process where the speaker and the listener both take part.

Furthermore, the speaker and the listener must have a *shared* understanding of the meaning of the words and sounds, gestures, expressions, symbols and context used in the communication. The key to effective communication is when there is a shared understanding of intention and meaning; both sender and receiver share an understanding of the purpose, sense and significance of the message. A shared understanding doesn't necessarily guarantee a shared agreement about what actions will follow from the communication – but it certainly makes it more likely.

Is it really that simple? Of course not. Why? Because communication is not a predictable, one-way event. It's a dynamic process, influenced by all the complexities and uncertainties of human behaviour.

Learnt or innate?

You are born with the ability to communicate. You have learnt to communicate by:

- listening to and watching other people
- being taught or trained
- practising and adjusting how and what you communicate according to the response you receive.

You have also learnt that there are basic *rules* of reciprocal communication. You have learnt that:

- people have to take turns to talk
- you have to wait until the other person has finished talking before you can speak
- the second person to speak should respond to what the first

person has said, rather than introducing a completely new subject.

Every day, you interact with other people who have different opinions, beliefs and needs from you. Your ability to exchange ideas and opinions with other people, understand their perspectives and solve problems between you will depend on how effectively you are able to communicate. Although your ability to communicate is innate, it is also a skill that can be learnt, developed and improved. You *can* become a skilled communicator!

The three elements of communication

Communication involves verbal, paraverbal and non-verbal elements:

1. verbal messages
2. paraverbal messages
3. non-verbal messages.

To communicate effectively, you must use all three elements to do two things:

1. send clear messages
2. receive and understand messages.

Let's look at the three elements of communication in more detail.

Verbal communication

Verbal communication is communication that uses words – either written or spoken – to convey thoughts, ideas and feelings.

Paralanguage

Paralanguage is used to modify meaning and convey attitude and emotion using such devices as pitch, volume, rhythm, intonation and emphasis.

Paralanguage gives communication its personality; the distinctive characteristics that make it unique.

All spoken communication has paralinguistic properties that are affected by emotions and attitudes. Attitudes (the way a person thinks and behaves) tend to be expressed intentionally and emotions (the way a person feels) unintentionally. If, for example, you are feeling angry or excited, your speech is likely to become rapid and higher pitched. When you are feeling defensive, your speech may be abrupt. When you are bored or feeling down, your speech probably slows down and takes on a monotone quality. Of course, attempts to fake or hide emotions are not unusual.

If you are sneering and sarcastic, your intonation and emphasis will leave the listener in no doubt as to your attitude!

Written communication can also include paralinguistic elements. These include punctuation, italics, capitalisation and the use of emoticons (facial expressions pictorially represented by punctuation and letters, such as :–) usually to express a writer's mood).

Paralanguage in written communication is limited in comparison with face-to-face conversation and, as you've probably discovered, this can lead to misunderstandings.

brilliant example

Paraverbal communication refers to *how* you say something, not *what* you say. A sentence can convey entirely different meanings depending on the emphasis on words and the tone of voice used. For example, the statement 'I didn't say you were stupid' has six different meanings, depending on which word is emphasised.

'I didn't *say* you were stupid.'

'I didn't say *you* were stupid.'

'I didn't say you were *stupid*.'

You can work out the other three meanings for yourself. (And, if you can't, show this page to someone else and see if they can work it out.)

Non-verbal communication

Non-verbal communication involves exchanging ideas, information and feelings without the use of words. This includes body language: posture, gestures and facial expressions.

Non-verbal communication provides clues to the attitude and feelings behind the words a person is using.

Although non-verbal communication usually reinforces verbal communication – you often send and interpret such signals subconsciously – it can also be used on its own to communicate something. For example, a look or a gentle touch can communicate a message clearly, as can a push or a slap. A smile or a frown can convey a message without the need for words.

Effective communication draws on all three elements – verbal, paralanguage and non-verbal – to exchange thoughts, ideas, opinions and feelings.

---

brilliant impact

The 7%–38%–55% rule

You may have come across the claim that communication is only 7% what you say; that communication comprises 55% body language, 38% tone of voice and 7% words. Not true!

Professor Albert Mehrabian, whose book *Silent Messages: Implicit Communication of Emotions and Attitudes* is the source of these statistics, told me that this is a misunderstanding of the findings. 'My percentage numbers apply only when a person is communicating about emotions and definitely do not apply to communication in general.'

So, it would be more accurate to suggest that the expression of *attitudes and feelings* is 7% what is said, 38% tone of voice and 55% body language.

Communication styles

Becoming aware of your communication style and those of other people is a good place to start when you want to improve your communication skills.

Each person has a unique way of communicating. Listen to your own speech. What sorts of words do you use? Which sort of body language and what tone of voice do you often use? In what situations and under what circumstances does your communication style change?

> each person has a unique way of communicating

Now, think of someone who you regard as a good communicator. Who do you know who can explain things clearly, who listens and understands what others are talking about? Someone you work with? A neighbour? A friend? Maybe it's a member of your family. There are great examples of communication everywhere – newsreaders, radio broadcasters, interviewers, politicians (at least, a few) and some celebrities. What is it about the way they communicate that you like? Compare your style to theirs.

Let's look more closely at your communication style. There are many communication models; that is, theories about the different ways that people communicate and the ways in which communication works. Knowing how to improve communication skills will become easier once you are aware of your own communication style; how you express yourself and how others perceive you.

The quiz below will help you to identify your communication approaches and attitudes.

Communication style quiz

Tick the statements that apply to you.

1	I often do more talking than listening.	A
2	I am more interested in facts than feelings.	T
3	If I get interrupted, I find it difficult to get back into the flow of what I was saying.	P
4	I often check to make sure I've understood what other people have said.	C
5	I prefer to talk about things rather than think about them.	A
6	I change the way I talk depending on whom I'm speaking to (for example, I speak more slowly and clearly with someone whose first language isn't English; I avoid using work-related jargon when talking with someone who doesn't work in the same type of job as I do).	C
7	I like to listen to information that will help me solve a problem or give me new ideas.	T
8	I can express my ideas clearly.	A
9	I like conversations and discussions to keep to the point.	P
10	I often have difficulty putting my thoughts or feelings into words.	T
11	I encourage other people to talk, and I ask appropriate questions.	C
12	When other people become emotional around me, I'm not sure how to react.	T
13	I use diagrams and charts to help express my ideas.	T
14	I often get so caught up in what I have to say that I am unaware of the reactions of my listeners.	A
15	Before I send a message, I think about the most relevant way to communicate it (in person, over the phone, in a note, email or text).	P
16	I like to make 'to do' lists and cross things off as I complete them.	P
17	I often do more listening than talking.	C
18	I enjoy conversations and discussions that take place at the same time as doing something else.	A
19	I take time to find the right words that will clearly express what I want to say.	T
20	I can tell when someone doesn't understand what I'm saying.	C

21	When talking with people, I pay attention to their body language.	C
22	I like meetings to follow an agenda and a timetable.	P
23	I will stop a speaker in mid-sentence if I disagree with a statement they have made.	A
24	If I don't understand something, I tend to keep it to myself and figure it out later.	T
25	I try to divert or end conversations that don't interest me.	A
26	To be really clear, I like to see things in writing.	P
27	I find it easy to see things from someone else's point of view.	C
28	I get straight to the point in emails.	P
29	If I find a conversation boring, I'll let my mind drift away.	A
30	My body language and gestures are quite controlled.	T
31	If I'm writing a formal letter or one with difficult or sad news, I often write it out several times before I send it.	C
32	If I have something relevant to add, I'll interrupt someone to make certain my views are heard.	A
33	I accept differences and conflict as a normal part of any work environment, and I know how to address them constructively.	P
34	I am completely at ease when a conversation shifts to the topic of feelings.	C
35	I try to anticipate and predict possible causes of confusion, and I deal with them up front.	P
36	I enjoy leading in a conversation (e.g. choosing the topic, controlling the pace).	A
37	I present my ideas so that others are receptive to my point of view.	T

Now add up how many of each letter you ticked.

- If you ticked mostly As you are an Active communicator.
- If you ticked mostly Cs you are a Connector.
- If you ticked mostly Ts you are a Theorist.
- If you ticked mostly Ps you are a Purposeful communicator.
- If (which is quite likely) you had similar scores for two or more communication styles, you use a combination of styles and communicate in different ways in different situations.

Active communicators

Active communicators communicate in a lively, spontaneous, opportunistic way. They are direct and straightforward; they get to the point quickly and keep things moving. They tend to make quick decisions and sum things up swiftly.

> active communicators tend to make quick decisions and sum things up swiftly

Active communicators like to talk about doing things and what they've achieved. They use language to make an impact and are prone to exaggerate.

They enjoy making people laugh and being made to laugh. They use large gestures and animated facial expressions and are not afraid to use strong language. They often like to move around during conversation. If you want a conversation with an active communicator you might have more success if you engage with them during a game of golf or tennis, rather than expect them to sit still and give you their full attention!

Active communicators like to be involved in discussion; they find it hard to sit on the sidelines unless they are amused or fascinated by what the other person is saying.

Whether they are familiar with what's being discussed or not, they come across confidently and persuasively.

Frustrations

Active communicators feel constrained by social rules and niceties, protocol and etiquette. They dislike hesitations and interruptions. If the other person pauses, they often take the opportunity to add their own comments or even change the subject. Active communicators may need to improve their listening skills and ability to empathise.

Connectors

These people like to take time to relate to and connect with others. They are friendly and approachable and welcome conversation with others. They show empathy and appreciation.

They particularly enjoy conversations where they feel they have made a connection which allows them to share personal aspects of each other's lives. They communicate in a way that will promote cooperation and harmony.

Connectors speak from the heart, are influenced by their feelings and emotions and use their intuition. They tend to be subjective (unlike Theorists, who are objective).

Connectors tend to listen more than talk – they like to 'read between the lines' and work out the feelings and intentions behind what another person is saying. They like to talk about relationships and people. They are interested in values and beliefs. They are usually optimistic and positive, and find conversations that help others feel positive and confident particularly rewarding.

They are tactful and considerate, generally thinking about how to phrase something so that it doesn't offend the other person.

Frustrations

Connectors dislike conversations that lead to conflict. They are upset by others who 'railroad' or 'talk down' to them. They prefer people to take time to connect with them and take their thoughts and feelings into account. They may need to be a little less sensitive and more assertive.

Theorists

People who use this style of communicating tend to think and deliberate during a conversation. They like to review what they've just heard and respond carefully in a way that most accurately expresses their thoughts on the matter. Their aim is to get information that will help them solve or discover something or provide new knowledge.

They tend to talk about ideas and information rather than feelings and relationships, and use wording such as 'I think' rather than 'I feel'. Unlike Connectors, they usually avoid emotional bonding and go straight for the facts. This approach can come across as quite cold and can be offputting to those seeking to first establish a personal bond.

Theorists enjoy talking about theories, ideas and strategies. During conversations they like to take in and develop new ideas and possibilities. They enjoy describing all aspects of a situation or an idea. Theorists are quick to pull together odd bits of information into rational arguments. They will persist with an argument if they feel that the evidence doesn't support what the other person has said.

theorists enjoy talking about theories, ideas and strategies

In a group conversation, they like to have time to think before they speak. Although they might not say a lot, inside their head is an active world of ideas and possibilities.

Frustrations

Theorists hate to look stupid; they want to appear capable and competent. They will consider what exactly to say to communicate their thoughts and feelings. They'd rather say nothing than say something wrong.

Theorists' body language is usually quite controlled and so may be hard for others to read. They could benefit from developing their group discussion skills, empathising and expressing and responding to feelings.

Purposeful communicators

Purposeful communicators like to be clear about the aim of a conversation, to stay on track and for everyone to remain respectful, with no interrupting! If they are interrupted and lose the thread of what they were saying, they are likely to start again from the beginning. They prefer others to complete their thoughts and finish what they are talking about before moving on to a different subject.

They like ideas and issues to be discussed in a logical order. Their language often uses extreme terms like 'always', 'never', 'must', 'should', or 'can't'.

Purposeful communicators prefer communication to be practical and realistic. They like to see things in writing so that they have something to guide them and refer to. They often keep 'to do' lists and enjoy the ritual of being able to achieve and then cross each item off their list.

They like to refer to past experience and conversations to see how these might inform their current situation.

Frustrations

Purposeful communicators don't like to waste time with chat and small talk, but would rather communicate with others in a way that will get things done. Purposeful communicators are curious to see if other people will follow through on what they say and get on with things.

They can be strongly opinionated and will speak out for and act on their own rights and the rights of others.

Purposeful communicators use 'proper' language and don't like to hear or use swearing, slang, offensive jokes or politically incorrect wording. They may come across as controlling.

Purposeful communicators may benefit from giving others time to express themselves, improving their group discussion skills and being able to discuss abstract ideas.

The introvert and extrovert continuum

A further way that communication styles can be understood is according to the extent to which a person's communication style is external or internal. Introverts (most obviously, Theorists) are motivated by their inner worlds while extroverts (most noticeably, Active communicators) are energised by outer worlds. So, introverts get their energy, inspiration and motivation from within – their own ideas, emotions and impressions. Extroverts get their energy and inspiration from external realities – other people and activities, and things outside of themselves.

Extroversion and introversion are usually viewed as a continuum. This means that one style of communicating might not be particularly different from another. Purposeful communicators, for example, appear to have both introvert and extrovert qualities. The extremes – the Active communicators and Theorists – are very different from each other.

It is thought that everyone has both an extroverted side and an introverted side, with one being more dominant than the other. But, of course, people shift in their behaviour all the time, and even extreme introverts and extroverts do not always act according to their type.

Communication styles – using your senses

The communication style quiz will have helped you to become more aware of the extent to which your communication style is introverted or extroverted. You may also find you have preferred ways of using your senses when communicating.

In the following quiz, for each question, circle the letter (a, b or c) that most applies to you.

1 If I have to learn how to do something, I learn best when I:

a watch someone show me how;

- b hear someone tell me how;
- c try to do it myself.

2 I check spellings by:

- a looking at the word to see if it looks correct;
- b sounding the word out in my head;
- c writing the word out in order to know if it feels right.

3 When I read, I often find that I:

- a visualise what I am reading in my mind's eye;
- b read out loud or hear the words inside my head;
- c fidget and try to 'feel' the content.

4 If I don't like someone, I feel uneasy as soon as I:

- a see them approach;
- b hear their voice;
- c sense they are nearby.

5 When asked to give directions, I:

- a see the actual places in my mind as I say them, draw them or give them a map;
- b explain them or write them down;
- c gesture, point, move my body with the directions I give, or in some cases go with the person.

6 When I have to remember a list of items, I remember them best if I:

- a write them down;
- b repeat them over and over to myself;

c move around and use my fingers to name each item.

7 When I choose food in a restaurant, I:

a choose from the descriptions in the menu;

b listen to the waiter or ask friends to recommend choices;

c look at what others are eating or look at pictures of each dish.

8 When I have to discuss something important with someone, I prefer to talk:

a face to face;

b on the phone;

c during another activity like walking or eating.

9 If I had a medical problem, I would prefer that the doctor explained by:

a showing me a diagram of what was wrong;

b describing what was wrong;

c using a plastic model to show what was wrong.

10 I am planning a holiday for a group of friends. I want to let them know what I have organised. I:

a use a map or website to show them the places;

b phone, text or email them;

c tell them where they'll be going and what they'll be doing.

Now add up how many a's, b's and c's you scored.

Mostly a's – You are a visual communicator.

Mostly b's – You are an auditory communicator.

Mostly c's – You are a kinaesthetic communicator.

Auditory communicators

Auditory communicators like to talk! They also like to listen. They enjoy discussions, talking things through and listening to what others have to say. Auditory communicators interpret the underlying meanings of speech through listening to tone of voice, pitch, speed and other nuances. Written information may have little meaning until it is actually heard. Auditory communicators often talk to themselves and repeat what they are told, either out loud or to themselves.

Visual communicators

Visual communicators translate what they say, hear and read into images and pictures. They pay more attention to talks and presentations if there is plenty of illustrative material, or if they can visualise or watch the other person demonstrate what they're talking about. When reading for study, they typically use colour highlighters. They often take notes by drawing pictures, diagrams or doodles.

visual communicators translate what they say, hear and read into images and pictures

Visual communicators easily remember faces and places by visualising them, and seldom get lost in new surroundings.

Kinaesthetic communicators

Kinaesthetic communicators tend to lose interest if there is little or no external stimulation or movement. They like to feel, touch and move. When they have to discuss something important with someone, they prefer to do this during another activity like walking or having a meal. When listening to talks and presentations, kinaesthetic communicators may want to take notes just for the sake of moving their hands. When reading for study, they typically underline and use colour highlighters. They often take notes by drawing pictures, diagrams or doodles.

Kinaesthetic communicators remember best what was done rather than what was said.

brilliant tip

Now that you are becoming more aware of your communication style, think about people you know – family, friends, people you work with – and how they communicate. Do any of the above communication styles help to describe and explain their way of communicating? Understanding how other people communicate is a key part of developing empathy, and one of the cornerstones of good communication.

There are many communication models but communication skills cannot be summed up in a few categories; remember, communication is a dynamic process, influenced by all the complications, strengths and limitations of human behaviour.

You may well find that you don't fall into one communication style but, like most people, have a combination of styles. You may also prefer one style of communication for one situation and a different one for another situation.

The aim is not to standardise the way you communicate – it's more about keeping your current communication style and adjusting to other styles of communication when necessary. Knowing that you have a different communication style from other people means that you can better understand any other differences between you, and adapt your communication style depending upon the situation.

brilliant tip

Get on the same wavelength

A good way to start adapting your communication style is to be more aware of someone else's. During a conversation, try to match something about the other person's body language and verbal style. Don't do everything they do (too weird!), but mirror just one thing.

For example, if the person gives mostly short answers to questions, ask shorter questions. Or, maybe they talk at a slower pace than you usually do. Slow your speaking speed to match theirs. This may sound overly simple, but it is a very effective way to become more aware of what and how other people communicate and get on the same wavelength.

Communication attitudes

> you have a communication *attitude* – a way of thinking and behaving which affects how you communicate

As well as being predominantly extrovert or introvert and using different senses to help you communicate, you will have a communication *attitude* – a way of thinking and behaving which affects how you communicate.

Assertive communication

Being assertive means expressing honestly and appropriately your feelings, opinions and needs. Assertive communication involves the belief that you can *choose* whether or not to tell other people what you think, how you feel and what you believe.

When you are being assertive you invite other people's views, even though they may be different from your own. You do not try to dominate others or involve yourself in criticism.

Aggressive communication

Aggressive communication also involves expressing your feelings, opinions and needs but in a way that threatens, dismisses or controls other people.

When you are aggressive you may feel you have to prove things and push a point. If you feel you are being treated badly you react with anger and hostility.

Rather than being honest and direct, aggressive communication often involves being rude, sarcastic and blaming.

Aggression is a one-way process – you say what you do and do not want but you do not listen to or take into consideration other people's needs and feelings.

Passive communication

Passive communication does not involve expressing thoughts, feelings and needs; it means letting other people talk over you and tell you what is and what is not. Passive communication means that you do not say what you think or feel. You often go along with others, even when you do not agree with them. You prefer to stay quiet and agreeable, but your compliance is frequently misinterpreted, often leaving others uncertain of your thoughts and feelings so they ignore or disregard you.

It's easy for other people to disrespect you.

Passive–aggressive communication

Passive–aggressive communication is an indirect and dishonest expression of feelings, opinions and needs. It is manipulative; you control situations and people without seeming to. Typically, this type of communication displays a passive resistance to cooperating with others.

Passive–aggressive communication often means suppressing anger and frustration and using a non-verbal way of expressing it – for example, giving others the silent treatment or dirty looks when you are unhappy with them. This does not, however, let others know what you are really feeling. Passive–aggressive communicators may also be in the habit of using sarcasm and other subtle communication devices to avoid confrontation or get out of doing certain tasks.

assertive communication brings together the key principles of excellent communication

In this book, the emphasis is on *assertive* communication. Assertive communication brings together the key principles of excellent communication; a shared process and one that promotes confidence and empathy between people.

Barriers to communication

Whatever your communication style, ask yourself whether how you communicate brings you closer to people or distances you.

Differences in the way people communicate can present barriers that get in the way of effective communication.

Dr Stella Ting-Toomey is a professor of Human Communication Studies at California State University. She has identified three broad cultural differences that can hinder effective communication.

Cultures provide people with ways of thinking about and interpreting the world. We all see the world differently, and the same words can mean different things to different people.

This can be attributed to a difference in culture (ideas, beliefs and values) between people, not just from different countries but from different parts of a country. Different organisations, generations and so on also hold differing values, beliefs and ways of behaving and communicating. So, as well as individual communication styles, each of us belongs to one or more cultures that have their own specific cultural communication styles.

Language barriers

The same words can mean different things to different people even when they appear to speak the same language. But if the languages are also different, and need translating, the potential for misunderstandings increases.

---

brilliant example

One of the more disturbing memories of the Cold War was when, in 1959, the Soviet leader Nikita Khrushchev said to the Americans at the United Nations: 'We will bury you!' This was taken to mean a threat of nuclear annihilation.

However, a more accurate interpretation of Khrushchev's words would have been: 'We will overtake you!' – meaning economic superiority. It was not just the language barrier but the fear and suspicion that the West had of the Soviet Union that led to the more alarmist and sinister interpretation.

Cognitive differences

Different cultures have different frames of reference and different ways of thinking about the world. These different world views provide a context for new information, knowledge and ideas to be compared, assessed and used. Cognitive barriers are created when the way people's thinking and reasoning give meaning to the world around them differs from another culture.

Behavioural differences

Each culture has its own ideas and beliefs about what is and is not appropriate behaviour, which affects verbal and non-verbal communication. This can be as simple as 'rules' about eye contact or how close you can stand to another person.

> each culture has its own ideas and beliefs about what is and is not appropriate behaviour

In some cultures, eye contact should be sporadic and people should stand at least three feet apart. In other cultures, eye contact may be considered disrespectful but it is acceptable to stand very close together.

Behaviour constraints can influence how much information one person gives another. In some cultures, discretion and restraint are important and it is considered correct to talk *indirectly* about an issue. In other cultures, people are more direct and say exactly what they mean.

Emotional differences

Different cultures have different ideas about the communication of feelings and emotions. Some cultures are generally open about their emotions, with hugs and kisses alternating with angry shouting and gesturing. They laugh and cry and are not afraid to show their anger, fear, frustration and other feelings. Other cultures strive to keep their emotions hidden and believe in communicating only the factual, objective aspects of a situation.

This can cause problems when people from different cultures communicate. One culture may think another is out of control emotionally. On the other hand, a culture that freely expresses its emotions and feelings might consider a more restrained culture to be uptight.

If the people involved are unaware of the potential for such cultural misunderstandings, they are even more likely to fall victim to them.

Personal barriers to communication

As well as cultural and personal communication styles, at any one time each of us may have personal difficulties which can create barriers to good communication. A range of negative emotions such as fear, anger, mistrust, jealousy, suspicion, distress, stress or anxiety can hinder communication and create misunderstandings. So can a lack confidence or feeling insecure or ignored. Discomfort caused, for example, by ill-health, pain, disability, sight impairment or hearing difficulties can also be a barrier.

Knowing about and wanting to understand differences and barriers is a key step towards more effective communication.

Confident communication

Understanding how differences in culture and communication styles can create communication difficulties can go a long way to helping you to develop empathy and understanding with other people. But even the best communicators can have their communication skills undermined when, for example, they are caught up in a whirl of emotions, face humiliation, feel misunderstood or are surprised by someone else's response. You can find yourself resorting to all kinds of inappropriate and unhelpful behaviour.

Being a brilliant communicator is often about feeling confident in situations where you may not feel comfortable.

Whether you're trying to lead a group of people, strike up a conversation at the bus stop, get someone else to calm down or tell someone you don't understand what they're talking about, there are two qualities that you need: *confidence* and *empathy*. Confidence means that you *believe* you have the ability to communicate well. Empathy gives you the ability to *understand* (but not necessarily agree with) other people's point of view, ideas, opinions and feelings.

being a brilliant communicator is about feeling confident

The good news is that if you don't think you have much confidence or empathy, you can learn to develop these qualities.

Confidence

First, it's important to understand that confidence is different from arrogance – that's when someone has an exaggerated opinion of their own importance. An arrogant communicator tries to impress others, while a confident communicator doesn't worry about whether others are impressed or not. Confident communicators are more concerned with mutual respect, consideration and understanding.

brilliant tip

Build up your confidence and self-esteem. Make a habit of talking to people everywhere you go. Say good morning to the person who delivers your post. Ask how their day is going.

If you're at the supermarket, talk to the checkout person. On the phone talking to someone at a call centre, ask how busy they have been today. At work, make a point of talking to someone you don't usually chat to. Acknowledge a neighbour and stop to exchange a few words.

The only way you'll condition yourself to talk to people and get over your fear of it is to do it frequently. What's the worst that can happen? Yes, you may be rebuffed or rejected, but feel the fear and do it anyway. By the time you've got over the fear, it'll be a habit. And good habits are hard to break!

Do confident people ever feel anxious about communicating with other people? Yes, they do, but the difference between them and people lacking in confidence is that effective communicators take action and take responsibility for the outcome of their communications.

Rather than focus on how much fear or anxiety they feel, confident people make use of their courage – they communicate with other people *despite* their fears or worries. They recognise they have to start somewhere.

Confident people take action – they don't wait to feel confident before they start communicating.

brilliant example

Ali met Tom when they were both working for the same company three years ago. Ali says, 'When I first got to know Tom, it was amazing how confident he was with other people. Whether he was conversing with the manager or colleagues, or negotiating with clients, he was able to communicate easily with anyone.

'More informally, he could chat with unfamiliar people about everyday topics and, within minutes, was joking and chatting with them like they were old friends.

'I envied Tom his lack of self-consciousness.

'The reason I didn't initiate conversations with people was because I didn't think they'd be interested in talking to me. Even though I had several interests, I thought that I wouldn't have anything interesting to say to someone, and that any conversation I tried to start would become awkward and strained.

'What I came to realise and believe is that *a lot of people* feel this way to some extent.

'In general, other people have good intentions – you have to start from a position of trust and believe that other people are not waiting to catch you out. If you are talking with them, they are simply focusing on you, the present situation and the topic of conversation. This realisation was a turning point for me!'

Simply knowing that other people often feel anxious and self-conscious about talking with and listening to others automatically gives you the second important quality needed for better communication skills: empathy.

Empathy

People with good communication skills want other people to feel positive about communicating with them. The most effective way to achieve that is to communicate with empathy.

Being empathic is different from being sycophantic; that's when a person is over attentive and their aim is to gain favour or ingratiate themselves with the other person. Being empathic simply means that you are willing to try to recognise and understand someone else's situation, point of view, thoughts and feelings.

This does not mean you have to agree with their ideas, opinions or feelings. It also does not mean that you make their situation your own. You can show interest in another person's situation and readiness to respond to their needs and feelings without taking over.

Empathic people are good at drawing on their own experience in order to relate to what others are saying and feeling. But they use it only as a starting point, not as the end itself.

Communicating with empathy means you accept that there are differences between you but you don't let those differences hinder communication. And, because empathy helps you to understand a person's experience from within their frame of reference, it goes a long way to overcoming cultural differences in communication.

empathy builds trust and can make communication easier and quicker

Empathy connects people. Empathy builds trust and can make communication easier and quicker.

brilliant tip

Imagine that a colleague, friend or family member calls and complains endlessly in a self-pitying way, and is not being open to solutions. Without empathy, it's easy to lose patience and sympathy. But put yourself in their shoes. You can do this by imagining what feelings and emotions they are conveying – stress, feeling overwhelmed, ill or victimised? If you're not sure, ask, 'Are you feeling stressed?' or 'Are you very worried?' Whatever they're feeling, you have probably felt the same, so you *can* empathise. Then you can continue communicating with that understanding in mind.

brilliant tip

Looking for clues

The ability to put yourself in someone else's shoes requires imagination and effort on your part. You have to do most of the work; other people don't always spell out what they're feeling. Others *do* give clues, not overtly, but in more subtle ways. Look for these clues. You have plenty to help you: what the other person says, what they *don't* say, how they say it and what they do.

Practise by being more aware of the connections between verbal language, paralanguage and body language. Do they all 'say' the same thing? Also, watch out for changes in the way the other person communicates.

The value of empathy comes not just from understanding the other person's feelings and frame of reference, but also from how and what you communicate as a result.

Communication strengths

No matter what level of empathy and confidence you have, the best place to develop your communication skills is from a position of strength. Using your communication strengths will help to build your confidence. Why? Because confidence comes from knowing you can do something well, and when you use your strengths you *are* doing something well.

From the quizzes in the last chapter you will have identified some of your strengths. How can you be sure if one of the ways you communicate is a strength? If it feels natural to you and allows you to communicate well in a specific situation, it's a strength. Maybe you communicate in a direct and straightforward way; you're clear and concise, you get to the point quickly and keep things moving.

Or maybe you feel comfortable taking time to relate to and connect with others, or you do well in conversations that consider and expand on new ideas and possibilities.

Perhaps you enjoy communicating with others during shared activities or you are good at taking in complex information just by listening. Maybe you find it easy to describe ideas and concepts to others simply by visualising the ideas as you talk.

brilliant **action:** Communication strengths

Go back to the quizzes in Chapter 1 and pick out three of your communication strengths. Write down examples of how and when you use those strengths:

- at work
- with family
- with friends
- in any other situations.

profanity
informal language
soft spoken

At work

C
C
T
T
T
P
C
C
T
C
P
A
T
C
C
P

C – 7
T – 5
P – 3
A – 1

Be aware of those areas where you are strong, where you have effective communication skills, and make use of them. The trick is knowing when to use which strength. Think of situations in which you've used your communication strengths successfully and times when you've used them unsuccessfully. Are there any patterns?

be aware of your strengths and make those your pillars

Be aware of your strengths and make those your pillars. That's when you are a confident communicator – you *know* you are able to communicate well and can apply and adapt your communication strengths depending on the other person and the situation.

brilliant example

Using communication strengths

Bryony works for a website development company. She enjoys hill-walking at weekends. Bryony's strengths are that she takes time to relate to and connect with others, particularly during shared activities. She enjoys conversations that explore new ideas and possibilities; she finds it easy to describe ideas and concepts to others in a clear and concise way.

At work, Bryony's strengths are her patience with clients and her ability to help them to clarify what they do and don't want. She takes time to understand clients' ideas and needs, and explains to them how they can or cannot be used on their website.

In her spare time, Bryony loves hill-walking because it gives her the opportunity to get into conversations with different people, expand her social life and keep fit while on the move!

Communication weaknesses

Having brilliant communication skills is not, however, simply a matter of identifying and using your strengths.

Remember, communication is a dynamic process, subject to a range of personal and cultural contexts. Sometimes it's possible to be too direct or too passive. For example, maybe you like to have time to think before you speak – this is a strength when you want to be sure you explain something clearly, but it could make things difficult for you in job interviews where you're under pressure to think on your feet. Or perhaps you like conversations and discussions to keep to the point. This can be a strength if you are chairing a meeting but can be a problem if someone else is in charge and allows discussions to wander off the point.

In these sort of situations, a range of expectations and emotions can get in the way of effective communication.

Expectations and emotions

When expectations and emotions get in the way, strengths, confidence and empathy can fall away and you can find yourself resorting to a range of negative communication and behaviour. For example, sulking and ignoring the other person or yelling, using abusive language and refusing to cooperate.

Ask yourself if you have reasonable expectations about the way other people communicate. Do you, for example, think that if someone else knows you well they should 'know' what you mean?

---

brilliant example

Are you on the same wavelength?

Couples in a relationship may think they are clear about what they mean when they communicate with each other – many of us think we communicate better with our partners than with strangers.

But research by Savitsky et al has shown that, often, partners in a relationship communicate with each other no better than they do with strangers. Why is this? It would appear that when we have a close relationship with someone, we believe that there is more insight and understanding between us than there really is.

Conversely, when people talk to someone they don't know, they automatically provide more information because they don't have a 'closeness bias' with that person.

The researchers carried out further studies between people who were friends. Just like partners, friends overestimated their effectiveness in communicating with each other. Communication problems arose when people expected each other to have the same perspective. It didn't occur to them that the friend may not share the same information or their perspective.

Often, we believe that there is a right and wrong way for us to communicate with each other. Most of the time, we are completely unaware of how our expectations can create communication breakdowns, misunderstandings, conflicts and distrust.

Incomplete communication, in which expectations are not clearly spelled out, can be confusing and allows emotions to take over.

Emotions are messengers that communicate something to you; they are communicating feelings and can play a big part in influencing how you communicate.

emotions can play a big part in influencing how you communicate

The part of your brain that springs into action when you are emotional is different from the area of your brain that operates when you are analysing, rationalising or reasoning. This means that when you are emotional (angry, jealous, excited, etc.) you cannot think or problem-solve. It's as if a wall has come down and you are behind it.

The same thing happens to other people when they become emotional. In any conversation you have, it's as if you are speaking with the emotion rather than the person. This doesn't often turn out well!

Thinking of emotions as positive or negative can limit your ability to communicate. Emotions just exist; there is no good or bad, right or wrong emotion. They are only positive or negative if the resulting behaviour and communication are positive or negative.

The challenge is to recognise and acknowledge your emotions, understand how you are feeling, put your emotions to one side and allow the thinking, reasoning part of your brain to take over.

Emotional intelligence is the ability to understand and manage your emotions, intentions and responses, as well as those of other people. When you demonstrate emotional intelligence, you are an effective communicator because you are not ruled by your emotions; you don't allow your emotions to take over and do all the communicating. You allow your emotions to *inform* your communications, not take over.

When you use emotional intelligence, your understanding and experience of emotions helps you to be empathic; you can take into account how the other person's emotions can affect what and how they are communicating. You use your own experience as a guide, but keep in mind that the other person might feel or think differently from you in any given situation.

without an awareness and understanding of emotions, it's difficult to communicate well

Without an awareness and understanding of emotions, it's difficult to communicate well. The feelings of others will escape you unless you're familiar with your own emotions. The more aware you are of your own emotions, the easier it will be for you to understand and communicate with other people.

brilliant tip

The better you are at knowing and managing yourself and your emotions, the better you become at understanding other people's emotions. Improve your emotional intelligence!

First, identify what emotion you are feeling. Second, ask yourself the message the emotion is giving you. And third, allow the message to *inform* what you say and do.

For example, an emotional response might be, 'I'm furious! It's all your fault!'

Using your emotional intelligence, your response would be, 'I'm furious. I am so angry I can't think straight. Let's talk about this when I've calmed down.'

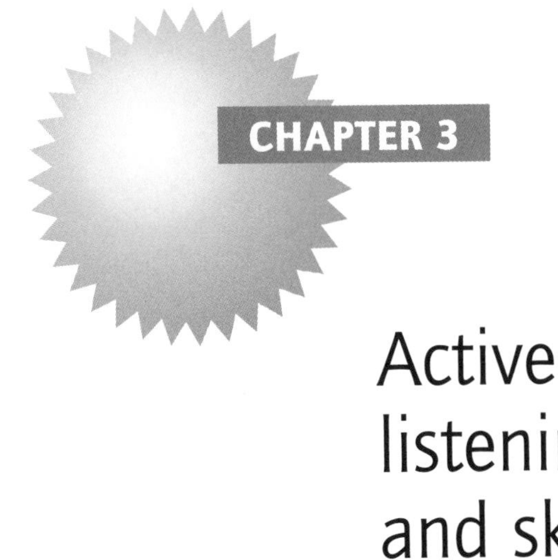

Active listening and skilful questioning

Every person I work with knows something better than me. My job is to listen long enough to find it and use it.

Jack Nicholson

Active listening

What, do you think, is the difference between listening and hearing? Are they the same thing?

If you hear something, you are simply *aware* of sound. Hearing is a *passive* process; you don't have to make any effort to receive the sounds. And the sounds you hear have no significance until you give them some meaning.

Listening, on the other hand, is an *active* process, requiring you to pay attention, interpret and derive meaning from the messages you hear and see.

Active listening takes things one step further. It is a communication technique, a structured form of listening and responding that focuses attention on the speaker with the aim of developing mutual empathy and understanding.

Active listening is a technique used in professional situations such as counselling, mediating, managing and mentoring. But the principles are useful for all of us in a variety of situations, whether it's discussing ideas at work, interacting with your children or making small talk in the supermarket queue.

The way to become a better listener is to learn and practise active listening.

How to be an active listener

Active listening is certainly not complicated. You simply need a positive, engaged attitude and lots of practice!

> there are two elements to active listening: acceptance and acknowledgement

There are two elements to active listening: acceptance and acknowledgement. They both help you to experience interest and understanding in what the other person is saying. Acceptance and acknowledgement also ensure that the other person *knows* you are interested and understand what they are saying.

Acceptance is a *passive* act; it is the ability to listen without attempting to negate what the other person is saying by, for example, interrupting, judging or dismissing. Communicating acceptance between you creates feelings of confidence and trust.

Acknowledgement is an *active* process; it happens in communication when you respond with recognition and consideration of what the other person has expressed.

Both acceptance and acknowledgement are conveyed using verbal and non-verbal communication.

Non-verbal responses

Non-verbal acceptance and acknowledgement can be as simple as making eye contact, a nod of the head, a smile or a touch on the arm. It can also be profound: holding someone, a sympathetic look or a gentle touch can reach out in ways that words cannot. Using facial expressions that reflect the feelings and content of what the other person is saying is also an effective way to show that you are engaged with what they are saying.

When someone is talking to you, they will (usually unintentionally) look for verbal and non-verbal responses to know whether or not they are being listened to. Non-verbal responses do not necessarily have to convey that you agree with the other person; you are simply showing that you are listening.

brilliant tip

The next time you talk to someone, be aware of *their* non-verbal responses. What do you notice they do to convey non-verbal acceptance and acknowledgement?

Verbal and paraverbal responses

Helpful responses facilitate communication by helping the speaker to feel understood. The brilliant impact box below sets out some of the key listening techniques, their purpose and examples. They are techniques for you to use when you are communicating with and listening to other people. You will be familiar with all of them – they're not difficult to recognise or understand – but the challenge lies in using them effectively in your relationships with other people. You'll need practice and patience!

brilliant impact

Listening technique	Purpose	Examples
Utterances	To show you are listening and are interested To encourage the speaker to continue	Uh-huh I see Uh? Really? Yes Oh

Listening technique	Purpose	Examples
Reflecting, paraphrasing, summarising	To check meaning To show you are listening and attempting to understand. Helps develop empathy	So, as I understand it … What I think you're saying is … Am I right? You think that …? You feel that …?
Clarifying	To avoid any misunderstanding and be clear	I just want to make sure I understand you. Can I clarify? Can I ask …? Can you tell me …? What did you mean by …?

Verbal acknowledgement and acceptance shows that you have been listening; that you are paying attention and focusing on understanding what the other person is saying.

Utterances. By using small verbal comments like 'yes', 'uh-huh', 'go on' or 'I see', you are signalling your interest and encouraging the speaker to continue.

Reflecting, paraphrasing. This involves reflecting on what the listener has said but using your own words and phrases, interpretation and understanding – often a summary of what the other person has said – to check you've understood their sense and significance. This can be quite a challenge; while the other person is speaking, you, the listener, have to keep a mental note of their main points or message. (You may use your senses to help with this – visual communicators, for example, may visualise the main points the speaker is making.)

When reflecting and paraphrasing, you do not have to agree with the speaker; you only state what you think the speaker said and how you have interpreted what was said. This helps each of you to know whether or not you've understood. If you haven't understood what the speaker is trying to express, they can explain some more.

Active listening is possibly the most important communication skill that you could learn – it's the quickest route to developing rapport, empathy and understanding. It is a skill which counsellors, negotiators, managers, sales people and teachers use more than any other skill. Of course, professional situations are a structured, organised way of communicating. It would be unnatural to reflect and paraphrase every time someone spoke to you.

active listening is possibly the most important communication skill that you could learn

However, active listening can be used in a wide range of formal and informal situations; the trick is to listen *as if* you were going to reflect and paraphrase (whether you do so or not). This is why *active listening* is so powerful. It helps you to really listen.

So, practise active listening and remind yourself that your goal is to understand what the other person is saying.

brilliant example

Active listening

Lin: 'I had an argument with Max at work and we haven't spoken since then.'

Chris: 'Really?' (Utterance)

Lin: 'Yes. It's been two weeks and it's not just bothering me, it's also causing a real atmosphere in the office. Today Jan, who sits opposite me, told me that they'd all noticed something was wrong between me and Max.'

Chris: 'Hmmm, it's affecting you and everyone else.' (Paraphrasing)

Lin: 'Yes, I'm quite upset. It's going to have to be sorted out, otherwise it'll create big problems when we all go away to the annual conference in July. There's quite a bit of planning that needs to be done for the presentation our team are going to be giving – both Max and I are involved in contributing to that. And, of course, there will be a party on the last night of the conference. That won't be much fun if we're still not talking. I really want to clear the air but he just makes me so angry, assuming that because I don't have children I can't possibly understand how hard it is for him to take on extra work. I have commitments too, you know.'

Chris: 'Let me get this right. You were angry because Max *thinks* you don't understand what it is like to have kids. Are you saying you're going to have to patch up your differences, otherwise it's going to make things difficult for planning and attending the conference?' (Reflecting, paraphrasing and clarifying)

Lin: 'Yes. Well, we don't *have* to sort it out, but things will be a lot easier for me, Max and everyone else if we do.'

Chris: 'What's the next step?' (Open question – see next section)

Lin: 'Maybe I should just tell him that I *do* understand it is hard, but that if we work closely on this, I'm happy to do some work this weekend so we can get ahead with the conference planning.'

Chris: 'Sounds good.'

Lin: 'Yes, that's what I think I will do.'

Actively listening makes the speaker feel understood and encourages open conversation.

Often, you'll find that you can interpret the speaker's words in terms of feelings. So, instead of just repeating what happened, you might add, 'It sounds like you felt *angry* when … happened.' The speaker might agree or adjust their message, for example, by replying, 'Not really angry, just confused.'

> When you are listening to somebody, completely, attentively, then you are listening not only to the words, but also to the feeling of what is being conveyed, to the whole of it, not part of it.
>
> Jiddu Krishnamurti

brilliant tip

Listening is an acquired skill; the more you practise listening, the better you get at it.

Listen to a discussion programme on the TV or radio. Pick out something someone says and practise ways to frame and re-frame sentences in various ways.

brilliant action: Practise active listening with a friend

One of you talk for two minutes on one of the subjects below. The other person must use active listening techniques to show interest and understanding.

Most importantly, when the speaker has finished speaking, the listener must reflect back what the speaker said and felt.

- The best job or holiday you ever had.
- The worst job or holiday you ever had.
- A pet you once had.
- What you'd do if you won a million pounds.
- What you like or dislike about Christmas.
- Other people's irritating habits.

Clarifying communication

So, to improve your listening skills, accept and acknowledge what the other person is saying. Listen without interrupting, and use verbal and non-verbal messages to show interest and understanding. Simple and straightforward? Well, yes, and no.

For a variety of reasons you still might not be clear about what the other person is telling you.

asking the right questions at the right time is at the heart of effective communication

A further way to increase understanding is, of course, by asking questions. Asking the right questions at the right time is at the heart of effective communication.

Here are some common questioning techniques, and when (and when not) to use them.

Open and closed questions

An effective way to clarify or get new information is to ask questions that encourage the other person to open up and explain some more. Asking 'open' questions does this. These encourage longer answers. But before we look at open questions, let's look at 'closed' questions. These are questions that require a yes/no answer.

Closed questions

Although asking closed questions can lead conversation down a dead end, closed questions are good for:

- obtaining facts
- being quick and easy to answer
- keeping control of the conversation with the questioner.

Closed questions are useful to establish facts and get a straightforward answer: 'Is this your pen?' 'Did you say tea or coffee'? But they can also be used for more deceptive purposes. Closed questions can be used to lead the other person to your way of thinking. They can do this in several ways. First, by framing ideas and opinions as rhetorical questions. Rhetorical questions aren't really questions at all, in that they don't require an answer. They are a persuasive technique, used to lead the other person to your way of thinking. Rhetorical questions are statements phrased in question form: 'If you stay out late tonight, what will happen in your driving test tomorrow?' 'Do you agree that we need to save the whales'?

Rhetorical questions can be effective in engaging the listener; they are more likely to agree with you than if they were simply told something: 'We need to save the whales.'

The first or last words of the sentence set up the dynamic of the closed, rhetorical question, indicating the quick and easy answer: words such as *do*, *would*, *are*, *will*, *if*. For example, 'Would you like to save some money?' or 'Are we all agreed this is the right course of action?'

Any opinion can be turned into a closed question that forces a yes or no answer (and solicits agreement) by adding tag phrases at the end of the sentence, such as 'isn't it?', 'don't you?' or 'won't they?'. For example, 'The cheaper option is better, *isn't it*?' or 'You all agree, *don't you*?'

Another way that a closed question can be used to manipulate the other person's thinking is with an assumption: 'If you were thinking of waiting until next year to buy this product, you need to ask yourself, how much more will it cost then?' This assumes that prices *will* go up next year.

Closed, leading questions can be framed in such a way that they are both quick and easy to answer and give you the answer you want, while leaving the other person assuming they've had a choice.

For example, 'Do you want to eat the broccoli before you eat your sausages, or after you eat your sausages?' (Useful for getting young children to comply!)

You must use leading questions carefully, though. If you use them in a self-serving way or in a way that undermines the interests of the other person, you are being manipulative and dishonest.

On the other hand, closed, leading questions can be used positively; when managing, mentoring or coaching, for instance. Closed questions can help get someone else to consider and agree with your suggestions. For example, 'Wouldn't it be great to get some qualifications?'

brilliant tip

Rhetorical questions are even more powerful if you use a string of them. 'Would you like to get some qualifications?' 'Qualifications that lead to an interesting, fulfilling job?' 'Would you like that job to be well paid, involve international travel and plenty of perks?'

Open questions

Closed questions elicit closed answers and keep conversational control with the speaker. Open questions do just the opposite; they invite open answers and hand control to the listener. Use open questions and you open up the dialogue and encourage the other person to talk.

Open questions ask the speaker for their knowledge, opinions, ideas or feelings. They usually begin with *what*, *why*, *how*, *tell me*, *explain* or *describe*. For example, 'Why do you think he said that?' 'How did that happen?' 'Tell me about your plans.'

Open questions are good for:

- getting specific details and general information
- obtaining ideas and opinions, and understanding feelings
- encouraging further conversation and giving control to the speaker.

Open questions can give you specific details. Maybe there

are gaps in what the other person has said (they may or may not be deliberately withholding information). Perhaps you want to check where the other person's information has come from,or you're not sure why they said something: 'How do you know …?'

Sometimes, if the other person is expressing a claim, theory or idea that you're finding difficult to grasp you may ask for specific examples. 'I'm not sure what you mean. Could you give me an example?' 'How does that work in practice?'

If the other person is wandering off the subject, you can ask, 'When you said … can you tell me how that relates to …?' You can also repeat what they have said with emphasis on the area where you want more detail. 'He *didn't* say you were stupid?' or, 'He didn't say you were *stupid*?'

Most importantly, if you want to really open up a dialogue and give control to the other person, open questions allow you to ask about the other person's thoughts, feelings, ideas or opinions: 'And how did you feel about that?' 'Could you tell me more about that?' 'Can you tell me what happened?' 'What did you think about that?'

Open questions can be asked whenever you need to clarify a point or find out more, at different points in a conversation. Open questions can also be used in a specific technique known as 'funnel questioning'.

Funnel questions

Funnel questions are a series of questions that seek further information that either goes into more detail or becomes more general.

Funnel questions that increase detail give you, the listener, more information about fewer topics. This technique starts with general questions, and then narrows in on a point in each answer, gathering more and more detail with each question. This guides the speaker and helps them to focus and recall detail.

For example:

Carrie: Tell me more about the meeting.

Sam: I think it went quite well. There were just a couple of problems.

Carrie: What, specifically, was one of the problems?

Sam: We couldn't agree a date for completion.

Carrie: What date, exactly, did she suggest?

Sam: The 15th of next month.

Starting by asking 'tell me more' is a general, open question that also focuses the other person on a chosen area, giving more information about this and giving you, the listener, the opportunity to pick up on and ask further questions about the specific details.

Using focus words like 'specifically', 'exactly' or 'particularly' directs the speaker to explain a particular point in more detail. Use these along with 'what', 'how' and 'when'.

For example, 'You said that she wasn't happy about leaving it to the end of the month. What, *specifically*, did she say?'

using the funnel questioning technique can help to defuse a potential conflict or heated situation

Using the funnel questioning technique can help to defuse a potential conflict or heated situation and help the other person calm down and feel understood. Funnel questions get them to go into more detail about their problem. This will distract them from their emotions, help you understand them and, possibly, know what to do to help.

brilliant tip

Listen to the radio programme *The Choice* on BBC Radio 4 to hear how funnel questions work.

Funnel questions that *increase* detail gives you, the listener, more information about fewer topics. In contrast, funnel questions that *decrease* the demand for detail broaden out the questions to give you wider information about more general topics. Ask questions that begin, for example, with 'who else' and 'what else', such as, 'So, she accepted the end of the month as a completion date. *What else* did you discuss?'

This style of questioning can be useful in situations where you want to encourage the person you're speaking with to open up. It can also be used to increase their confidence.

brilliant example

Don't be afraid to ask questions. Asking questions does not make you look stupid. Listen to news programmes on the radio and TV. Note how often John Humphries on Radio 4's *Today* programme and Jeremy Paxman on BBC 2's *Newsnight* ask questions to clarify both their and the listeners' understanding.

brilliant tip

Remember to be aware of your non-verbal communication when you are asking questions. Be sure that your questions don't come across as interrogative, attacking, defensive or rude. Your body language and tone of voice all play a part in the answers you get when you ask questions.

brilliant tip

Make sure that you give the person you're questioning enough time to respond. They may need to think before they answer, so don't interpret a pause as a 'no comment' or an opportunity for you to take over the conversation.

The benefits of active listening

Active listening is a powerful technique that will enable you to achieve the following:

Concentrate on what the other person is saying. Active listening is intrinsically focused.

Whether or not you reflect back to the other person what they have said, because you are listening *as if* you were going to summarise what the speaker is saying, active listening compels you to concentrate your attention. And, because you are focused on the other person, you will find that you are not distracted by whatever else may be going on around you. Active listening prevents you from being distracted by external circumstances or internal issues.

It helps you avoid thinking about what you are going to say next (particularly helpful in conflict situations or disagreements).

Increase your understanding. Active listening helps you to interpret what the other person has said, in your own words. Reflecting back and asking relevant questions allows the speaker to confirm, adjust or correct you if you have misunderstood.

Overcome your assumptions. Because everyone has different perspectives, the speaker and the listener may attach different meanings to the same statement. Your assumptions, emotions, judgements and beliefs can distort what you hear. Active listening techniques can overcome those barriers because, again, the speaker can confirm or refute your interpretation and help you see things from their point of view.

Develop empathy and rapport. Active listening and empathy are mutually inclusive. By trying to understand what the other person is saying and feeling, you are trying to see things from their point of view.

Increase the speaker's understanding and encourage them to open up. When you reflect, paraphrase and summarise what you have heard, you help the speaker reflect, then confirm, retract or adjust what they meant. This can also encourages them to open up and say more.

Improve your ability to influence, persuade and negotiate. Active listening increases your understanding of the speaker's intentions, feelings and motivations. So you are more likely to make the right response and say whatever is appropriate, to strengthen negotiation, diplomacy and cooperation.

Be more likely to remember what was said later. You retain less when you engage in mindless listening. Active listening is mindful listening.

active listening is mindful listening

Avoid being bored. Active listening helps you to *engage* with what the speaker is saying; to look for points of potential interest.

Deter you from interrupting. You don't interrupt and break into what the other person is saying with unnecessary questions or comment. You recognise that interrupting is a waste of time; it frustrates the speaker and limits full understanding of the message. (However, there can be good reasons for interrupting, such as a show of support or enthusiasm, or asking for clarification.)

Allow the speaker to vent feelings. Because you don't interrupt!

Give you control in difficult situations. Reflecting back slows everything down. This gives both sides time to think.

brilliant tip

Remember, a good listener tries to understand what the other person is saying. In the end you may disagree but, because you listened, you know exactly what it is you are disagreeing with!

Use your preferred communication style. If you're an Active communicator, you prefer to talk and do things more than listen and sit still. Active listening gives you something to do! Active listening is an active process.

If you are a Theorist, active listening helps you to clarify other people's ideas, understand their points of view and take feelings into account.

For Purposeful communicators, active listening gives you a purpose; the purpose being to improve your understanding.

Finally, although Connectors are already effective listeners, active listening can further increase the skills you have of relating to, understanding and connecting with other people.

brilliant tip

What happens if you can't listen? If you are too busy, distracted, confused or worried to focus on what the speaker is saying? Say so! Explain that this isn't a good time for you. Tell the other person or negotiate a time when you will be better able to give your full attention.

brilliant recap

- Show interest – use verbal and non-verbal messages to accept and acknowledge what the other person is saying.
- Check if you have understood; reflect, paraphrase and/or summarise key points to confirm your understanding.
- Ask questions to increase understanding and dialogue.

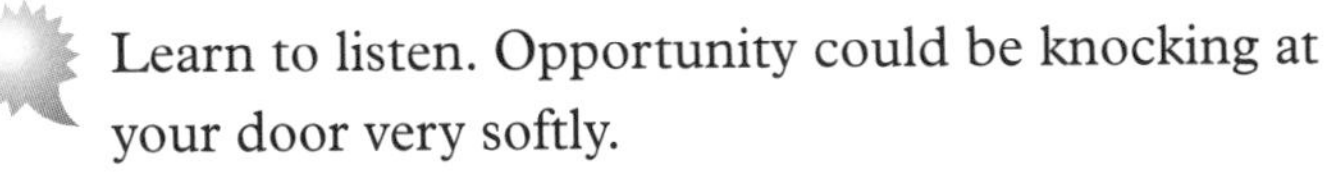

Learn to listen. Opportunity could be knocking at your door very softly.

Frank Tyger

Reading between the lines: feelings, emotions and attitudes

Although claims that communication is 93% non-verbal communication and only 7% words are inaccurate, if you have ever watched TV with the sound turned off, you know that it's not too difficult to pick up the essence of what people are saying and feeling simply by paying attention to their non-verbal communication.

brilliant recap

The 7%–38%–55% rule

The claim that communication is 55% body language, 38% tone of voice and 7% words is inaccurate. Professor Albert Mehrabian carried out research on verbal and non-verbal messages, feelings and attitudes. He has stated that this is a misunderstanding of the findings and that these percentages apply only when a person is communicating about emotions, not communication in general. So it would be more accurate to suggest that the expression of *attitudes and feelings* is 7% what is said, 38% tone of voice and 55% body language.

What is beyond question is that non-verbal behaviour is a powerful communicator. Often, people immediately and most emotionally respond to the non-verbal aspect of communication.

I speak two languages. Body and English.

Mae West

In Chapter 1 you will have read that although non-verbal communication usually supports, moderates or emphasises verbal communication (speech-dependent), it can also be used on its own (speech-independent) to communicate attitudes, emotions and feelings.

Let's look at the science. Non-verbal communication begins in your limbic brain – the part of your brain that reacts instinctively to the world around you. It responds in an automatic and genuine way. It's also where your emotions spring from; emotions that occur spontaneously, without any rational thought or reasoning.

In fact, limbic (emotional) responses are hardwired into your system, which makes them difficult to hide, like trying to suppress a startle response when something makes you jump.

So, limbic responses are a genuine reflection of your feelings, attitudes and intentions and are often expressed non-verbally. You use the limbic part of your brain when you express yourself emotionally and/or react to other people and events.

limbic responses are a genuine reflection of your feelings, attitudes and intentions and are often expressed non-verbally

Another part of your brain, the neocortex – the new brain – is responsible for thinking, remembering and reasoning. It is this area that gives you the ability to evaluate and understand the thoughts and behaviour of both yourself and others.

Speech is mostly conscious and intentional. Typically, the neocortex formulates, controls and expresses your thoughts, ideas and opinions.

On the other hand, non-verbal communication (whether it accompanies and supports speech or is speech-independent) is mostly unconscious and unintentional.

Because you are usually unaware of how much you are conveying non-verbally, non-verbal messages often reveal your thoughts, feelings and emotions more genuinely than what you actually say.

In the same way, when you are communicating with other people, you are reading or picking up on *their* non-verbal communication without being aware of it. In fact, what some people might call their 'intuition' is actually this unconscious process of picking up and processing other people's non-verbal information.

Why do you need to improve your non-verbal communication?

Becoming more aware of non-verbal behaviour will greatly improve your communication skills. You will be able to do the following.

- Pick up on a range of emotions, gauge moods and be more likely to know when others are, for example, sad, lonely, happy, proud or disappointed. You'll be able to 'read' body language, and with enough practice it'll become second nature, like looking both ways when you cross the street.
- Increase your ability to understand other people's behaviour, anticipate problems and avoid hidden pitfalls.
- Know whether or not you are getting mixed messages and be able to identify contradictions between what is being said and what that other person is really thinking.
- Develop empathy and rapport more easily than if you just listen to what people say.
- Increase your ability to persuade, motivate and influence other people.

- Convey feelings that you're not actually experiencing. Supposing, for example, you are about to give a presentation, or attend a social event where you don't know anyone, and you're not feeling as confident as you'd like to be. Adopting non-verbal 'confidence' signs and signals will not only make you appear more confident, you actually will feel more confident!
- Ensure that what you say is consistent with how you feel.

? brilliant questions and answers

True or false?

1. If you are explaining something to someone and they cock their head, it means that they are confused.
2. If someone looks down at the floor a lot, they are probably shy or timid.
3. Crossing your arms means you're feeling defensive.
4. Lowered eyebrows and squinted eyes show that a person is attempting to understand what is being said or going on. It's usually sceptical.
5. A fast tapping movement of the foot signals impatience.
6. If a person talks quickly and mumbles, or isn't clear on what they are saying, they are lying.

Answer: While these non-verbal behaviours *can* indicate specific feelings and attitudes, body language is more subtle and less definitive than you might think. For example, if you are explaining something to someone and they cock their head, it could that they are confused. But it could also mean that they are challenging the speaker or that they have a crick in their neck!

Equally, if someone looks down at the floor a lot, they are probably shy or timid. Or maybe they are just checking out their new shoes! But they might come from a culture that believes that looking at someone in the eyes is a sign of disrespect.

There are times when you or the other person may send mixed messages – we say one thing yet our non-verbal communication reveals something different. This non-verbal language will affect how you act and react to others, and how they react to us.

Let's look at some of the many ways in which we all communicate non-verbally, so that you can use and understand these signs and signals to communicate more effectively.

Facial expressions

Facial expressions can say so much; think what can be conveyed with just a smile or a frown. Often, your facial expressions will reveal your true feelings. While you may say that you are feeling fine, the look on your face may tell people otherwise.

often, your facial expressions will reveal your true feelings

An inability to read facial expressions, which is a characteristic of people with the developmental disorder Asperger's syndrome, is a social disadvantage, making it difficult to communicate with and understand others. In fact, it may often be difficult for any of us to accurately determine the subtle differences in facial expressions of, for example, sad, depressed, jealous, bored or worried. Although some facial expressions and their meanings vary between cultures, facial expressions are also among the most universal forms of body language. Facial expressions for happiness, sadness, anger and fear are similar throughout the world.

Eyes

A person's eyes are thought of as the 'windows to their soul' since, to a greater or lesser extent, they are capable of revealing what you are feeling or thinking. Even, for example, if a person is angry and hiding it with polite words, their eyes can tell a different story.

When you are talking to another person, eye contact and facial expressions provide you with important social and emotional information. Without consciously doing so, you read eyes and faces for positive or negative mood signs.

The most obvious thing that comes to your attention is whether people are making direct eye contact or averting their gaze. Certainly, if someone is lying to you, they may avoid direct eye contact. On the other hand, they may overcompensate by looking you in the eye to assure you of their sincerity.

Remember, though, frequency and interpretation of eye contact vary between cultures.

Mouth

Although a smile is one of the clearest non-verbal signals, smiles can be interpreted in many ways. A smile may be genuine, but a smile can also convey sarcasm, scepticism or cynicism.

A mouth that is turned up or down can indicate a range of feelings. Pursed lips might be an indication of disapproval, distaste or distrust. People often cover their mouth to cover a smile or smirk and may bite their lips when they are worried, anxious or stressed.

brilliant example

A frown, a smile, a hand on the arm – these can convey so much. There is an often-quoted story about a man who committed suicide by jumping from the Golden Gate Bridge in San Francisco. He had left a note in his apartment that said he would not jump if, on the way to the bridge, he met one person who smiled at him.

A smile cannot, of course, save everyone; the motivations behind suicide are too varied. But such simple non-verbal communication can be surprisingly effective.

brilliant tip

Reading facial expressions

Start with the basics: what does anger, disgust or surprise look like? Is it the same in everyone?

Watch TV with the sound turned down or watch foreign movies to help you to develop your ability to read and understand facial expressions. Reality TV programmes are particularly useful for observing facial expressions and other non-verbal communications. Notice how the participants respond to different activities and situations. Because you're not distracted by what is being said, turning the sound down is a good way to 'see' emotions, attitudes and feelings.

This is good not just for understanding individuals' attitudes and feelings, but also for developing your understanding of group dynamics.

Gestures

Gestures can be some of the most direct and obvious non-verbal communication.

Although speech-*independent* gestures are used deliberately and specifically to replace words and have a direct verbal translation, they can have different interpretations in different cultural contexts, ranging from complimentary to highly offensive! Giving a thumbs-up or a peace sign might have completely different meanings in different cultures. The gesture, for example, made by touching together the thumb and index finger in a circle while extending the other three fingers is usually used to mean 'OK'. But in some parts of Europe, the same signal implies 'you are nothing'.

Speech-*dependent* gestures are those gestures used spontaneously when you speak.

They are integrally connected to speech and thought processes and are used to support or emphasise what is being said, for example, when a person points to what they are talking about or uses their fingers to indicate numerical amounts.

Posture

A person's body language and posture can convey a wealth of information about their level of attention, involvement and attitude (think of a bored, disengaged teenager!) and feelings towards another person.

Direction of lean, body orientation, arm position and how open or closed a person's body is all give clues to attitudes and feelings such as anger and hostility, shyness and discomfort.

With *open* postures, the body is open and exposed. This type of posture generally indicates a relaxed, calm, confident, engaged and approachable attitude.

Closed postures involve keeping the body obscured or hidden, often by hunching forward and keeping the arms and legs crossed. This type of posture usually indicates more negative feelings of hostility, stress or anxiety.

Haptic communication

Haptic communication is the way in which people communicate via touching. It is an important part of non-verbal communication in interpersonal relationships.

It can include, for example, a handshake, a pat on the back, a high five or a kiss. This type of non-verbal communication can reveal intentions or feelings; touching someone's arm can signal that you want to talk next or interrupt.

The meaning conveyed from touch is highly dependent upon the context of the situation, the relationship between communicators and socially acceptable levels of touching. In some cultures, for example, touching someone else's head may be offensive. In a recent study of groups of people communicating, it was found that in England, France and the Netherlands touching was rare compared to the level of touching among Italians and Greeks.

Proxemics

If you've ever felt uncomfortable when someone else has stood too close to you, then you'll know the impact that personal space can have on people and their ability to communicate.

The term 'proxemics' refers to the distance between people as they interact. Just as gestures, posture and touch can communicate a great deal of non-verbal information, so can the physical space between people.

The amount of distance we need and the amount of space we perceive as belonging to us is influenced by a number of factors including context, culture, personality characteristics and level of familiarity.

Below is an outline of four areas of personal space – public, social, personal and intimate – that we all (usually unconsciously) respect and use.

- *Intimate distance – 6 to 18 inches.* This level of distance between people usually indicates a close and comfortable relationship. It often occurs during intimate contact such as hugging, whispering or touching.
- *Personal distance – 1.5 to 4 feet.* Physical distance at this level usually occurs between people who are friends or family members, and to separate people waiting in queues.

- *Social distance – 4 to 12 feet.* This level of physical distance is often used between individuals who are acquaintances or strangers, in cases where you do not know the other person well, or in public areas – shops and bus stops, for example.
- *Public distance – 12 to 25 feet.* This is usually the distance maintained in a public speaking situation, between a speaker and the audience, for example, when talking to a class of students or giving a presentation at work.

As with all types of non-verbal communication, personal space is subject to cultural variances. In one culture, you might find yourself almost nose to nose with a business associate because their social space is equivalent to your intimate distance. But in another culture, you could find the other person backing away from *you*, trying to regain what they view as appropriate social space while you follow them across the room!

Making sense of non-verbal communication

Although there are broad categories and ways to interpret non-verbal communication, it's important to note that:

- it's not always possible to interpret a single gesture or signal in isolation
- although some facial expressions appear to be universal – for example, shock, disgust and joy – and need no interpretation, others are more subtle
- the meanings of non-verbal communication differ from one person to the next, and there is a wide range of cultural differences, too.

So, just how can you make sense of a person's body language, gestures, facial expressions and tone of voice? The key to understanding non-verbal communication is to register and understand the *context*; that is, the circumstances accompanying what is being said.

This means taking into consideration all the peripheral, seemingly unimportant, circumstances that are, in fact, entirely relevant. It is these supporting conditions that determine the meanings of the non-verbal messages we send and receive.

When it comes to non-verbal communication, context is all! But what contributes to context?

- Words and language.
- Motions and feelings.
- Culture, circumstances, conditions, situation, background.

Rather than interpret single gestures or facial expressions, the trick is to look for 'clusters'. Clusters are when a number of verbal and non-verbal communications and actions occur close together and so indicate a consistent message.

Clusters of non-verbal communications provide a much more reliable indication of meaning than one or two signals in isolation. A single body language signal isn't as reliable as several signals, so look out for several signs that all seem to be saying the same thing.

clusters of non-verbal communication provide a much more reliable indication of meaning

For example, you might notice someone across the room in a meeting who is continually looking at their watch, and conclude they are anxious to leave on time. Look again for more repeated actions such as tapping their pen, shifting around or slumping in their seat. Their face may also show a distinct lack of interest and appear blank.

It's more likely they're bored!

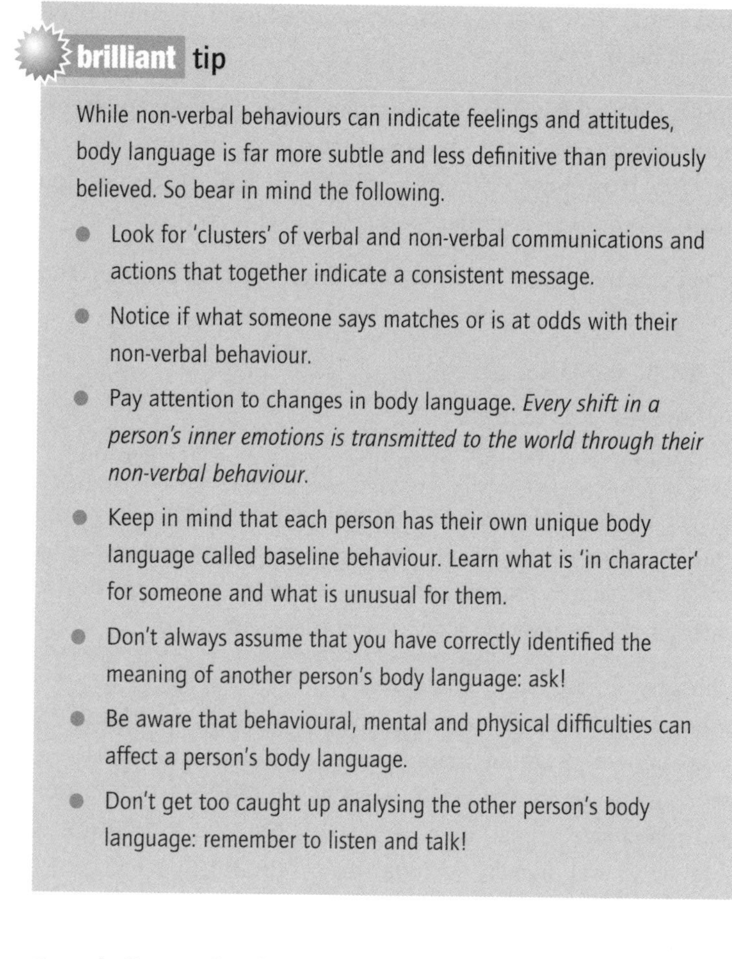

brilliant tip

While non-verbal behaviours can indicate feelings and attitudes, body language is far more subtle and less definitive than previously believed. So bear in mind the following.

- Look for 'clusters' of verbal and non-verbal communications and actions that together indicate a consistent message.
- Notice if what someone says matches or is at odds with their non-verbal behaviour.
- Pay attention to changes in body language. *Every shift in a person's inner emotions is transmitted to the world through their non-verbal behaviour.*
- Keep in mind that each person has their own unique body language called baseline behaviour. Learn what is 'in character' for someone and what is unusual for them.
- Don't always assume that you have correctly identified the meaning of another person's body language: ask!
- Be aware that behavioural, mental and physical difficulties can affect a person's body language.
- Don't get too caught up analysing the other person's body language: remember to listen and talk!

Speak fluent body language

One of the benefits of becoming more aware of and 'reading' other people's body language is that you become more aware of your own non-verbal messages. This can make a great difference to your ability to communicate effectively.

Your non-verbal communication can help you to emphasise or moderate what you say in words. Non-verbal behaviours communicate who you are and how you feel. Other people draw conclusions from your body language about your sincerity, credibility and emotions.

> non-verbal behaviours communicate who you are and how you feel

Even if you are silent you are still communicating through your posture, facial expressions and appearance. For example, a combination of minimal eye contact, slouching and fidgeting projects a negative attitude. On the other hand, a balanced posture, appropriate eye contact and calm gestures can help communicate a confident approach.

Does your non-verbal communication help or hinder your ability to be a brilliant communicator? Here are some non-verbal signs and signals to be aware of.

Posture

Think about how your perceptions of other people are affected by the way they sit, walk, stand up or hold their head. Some people come across as confident communicators before they've said a word. Why is this? More often than not, it's to do with their posture.

The way you hold yourself has an impact on how you communicate: remember, you want people to want to listen to you.

Not only is slouching and slumping bad for your health (you could end up with neck and back problems), poor posture also has a negative impact on how others perceive you. If you slouch, stuff your hands into your pockets, shuffle your feet and avoid eye contact, people will get the impression you don't want to communicate with them.

Your posture is often influenced by your emotions. If, for example, you are feeling intimidated or anxious, it will show: hunching and drooping give away that you are ill at ease, and the chances are that others may feel just as uncomfortable around you!

But even when you are feeling intimidated or self-conscious, you can convey feelings that you're not actually experiencing: you can positively influence your mood or emotions by simply changing your posture. To communicate confidence, simply stand (or sit) straight and keep your head level. Relax your shoulders and spread your weight evenly on both legs.

Try out your confident posture in front of the mirror and make yourself aware of what the 'confident you' looks and feels like. Then, when you are about to enter a situation where you are not as confident as you'd like to be, you can adopt these 'confidence' signs and signals to project confidence.

Distance and touch

Learn what the most comfortable distance is for you, in different situations. Allow yourself enough room to feel at ease and move when necessary. Look for clues about what is personally, socially and culturally acceptable in regard to touching other people. Know that, for example, touching a colleague on the arm or shoulder when they are upset is acceptable as an expression of sympathy. But doing the same thing when you are discussing a project you are working on together may come across as over familiar, patronising or even controlling!

Gestures

Do you use hand and arm gestures in nearly every sentence? Do they support or detract from what you are saying? Try to avoid a continuous physical interpretation of everything you say. Instead, add gestures selectively where they add the most impact. When you're talking, the natural action for your hands, for example, is for them to move in support of what you say. Try to use calm, flowing hand movements rather than short sharp movements, which can distract or intimidate the other person.

Eyes

Think about the last time you spoke to someone who continually avoided eye contact. They may have looked over your shoulder, at the floor, or even at someone else – everywhere but at you. It is most likely that you felt uneasy or frustrated and you doubted the other person's interest. Eye contact is also important in maintaining the flow of conversation and for gauging the other person's response.

There are times, though, when face-to-face conversations with a high level of eye contact can be intimidating and off-putting. There will often be times when it's more appropriate, relaxed and conducive to the other person opening up if the conversation takes place while you are involved in an activity alongside each other.

---

brilliant example

Tony, the father of two teenage boys, suggested they had more mealtimes together so that they could discuss their day, schoolwork, plans for the weekends and so on. The boys found the formality too intense. Mealtimes became strained and the boys said as little as possible.

Tony soon realised that his sons were far more likely to open up to him if they chatted while he was preparing the meal. The informality seemed to strike just the right notes: informal, natural and relaxed.

Paraverbal communication

> avoid gabbling – rapid and indistinct words confuse people and can result in your not being understood or taken seriously

Tone, pitch, volume, inflection, rhythm and the speed that you speak all have an impact on what you say. These non-verbal speech sounds provide subtle but powerful clues to your true feelings, attitude and intentions. Think about how tone of voice, for example, can indicate sarcasm, anger, affection or confidence.

Try to speak clearly and calmly. Avoid gabbling – rapid and indistinct words confuse people and can result in your not being understood or taken seriously.

Effective non-verbal communication

- Improving your posture and working to eliminate nervous quirks and mannerisms can take time and effort. But if you focus your energy on one thing at a time you'll quickly improve your overall non-verbal communication.
- Look at the big picture. Don't stay awake at night wondering if your little finger is playing its part in expressing what you want to communicate. You don't have to get every nuance 'right'. Remember, different parts of your body work together to communicate in clusters, so if you are being clear and honest it is more likely that your non-verbal communication will support what you say.
- Say what you mean and mean what you say. Verbal and non-verbal messages are co-expressive – if you say what you mean, your body language will follow.
- If you're not sure how you feel about something or someone, pay attention to what your body is saying. Just as other people can read your body language to help uncover what you're feeling, you can also learn from your body. Tense feelings in your stomach? Uneven breathing? What might that be telling you?
- If you think a sign or signal has been misinterpreted, say so: state your intended meaning right away. For example, if you cross your arms to keep warm, say 'I'm cold, are you?' That way the person won't think that you're closing off.

- It's advisable to adopt a balanced posture, calm voice and gestures to help you look and feel confident. But be aware that faking a gesture or facial expression can come across as phony and insincere.
- When people say that someone seems false, they're usually referring to mannerisms that seem faked; they don't truly match what the speaker is saying.
- Even when you deliberately try to control your body language, there is still what is known as 'leakage' that others can pick up. For example, your face may 'leak' information about how you feel; although you may try to appear welcoming, you may be unable to hide your embarrassment when meeting someone who has a facial deformity.
- Use the most positive (or, when necessary, negative) gestures and facial expressions first and last. Typically, you make your most memorable impressions within the first five to ten seconds. You also leave a lasting impression in the last five to ten seconds.

What to say, and how to say it

> There is only one rule to become a good talker, learn how to listen.
>
> Author unknown

Have you ever wondered, 'How am I going to explain this?' or 'What do I say next?' or 'Am I ever going to get a chance to say my bit?'

Polly has just been promoted to a supervisory role. 'Most of the time, I talk too much. I don't really think before I speak – I waffle on and use exaggerated gestures which some people find off-putting. I'm worried that I'm going to have difficulty explaining things clearly and concisely to the people in my new role at work.'

Rob panics when it comes to making small talk. 'Whether it's at a social occasion or trying to establish rapport with a client, either I don't know what to say or I'm afraid of saying the wrong thing.' (See the section on 'How to make small talk' in Chapter 7.)

Then there's Joe. He is interested in other people and likes to get to know them better. He's a good listener but, at work and in social situations, Joe often finds that he gets cornered by boring people. Joe doesn't want to offend them because they're harmless really. 'How do you say "shut up" nicely?' he wonders. (As in all good stories, the answer is at the end of this book.)

Lianne likes to keep conversations on track. If she's interrupted, she's likely to start again from the beginning. 'I find it difficult to keep up in meetings when the subject switches from one idea to another.'

Finally, there's Jim. Whether it's his teenage daughters or a colleague, Jim finds that, too often, they either don't understand or don't do what he asked. 'No matter how often I repeat myself, they still don't get it.'

you have to communicate in a way that makes it easy for other people to listen and to understand what you mean

There are no hard and fast rules about how to talk to other people. But, whether you are making small talk at a party, trying to disengage yourself from a boring person, or you need to get your point across clearly, there is one key principle: you have to communicate in a way that makes it easy for other people to listen and to understand what you mean.

This can involve:

- being clear about what and why you want to say something
- being aware of not just what you say, but also how you say it
- being aware that your perspective may differ from other people's
- understanding the other person's perspective and situation
- asking and inviting questions and feedback.

Set up the listening. Prepare who you are talking to for what you want them to hear. Get people to listen as a possibility rather than a problem.

Mal Pancoast

Let's look in more detail at how to make it easier for others to listen and understand.

Being clear and direct: get results through assertive speaking

Communicate purposefully. First, think about the overall message that you want to convey. If you are making small talk at a party, for example, you want to convey that you are friendly, approachable and open to conversation. If you are explaining to someone how to do something, your aim is to enable them to do it for themselves.

Think before you speak. It's perfectly acceptable to pause when it's your turn in a conversation and give yourself a moment to organise your thoughts and think about what you want to say.

Sometimes, when it comes to the words you use, the exact wording is not important. At other times it's crucial. For example, say 'I need this by next week' and the other person will assume any time next week is OK. Say exactly what you mean – 'I need it on Monday morning' – and you're more likely to get exactly what you want.

If you're not clear what your message is and you don't use the right words, it's all too easy for the other person to misinterpret what you say. And sometimes they'll take advantage of your ambiguity and misunderstand deliberately!

Vagueness is all too common in any number of situations – between parents and their children, between friends and between people at work. How often do you use all sorts of indirect ways to let others know something? Hinting, implying, using sarcasm and rambling are all *indirect* ways to say something; they all hide the meaning of what you *really* want to say. The best way to make it easy for others to listen and understand is to be clear and direct about what you say and mean.

brilliant tip

Say it in one

Hit the headline first. Think about the main point that you need to make. Put it in a sentence. Then elaborate, following the newsreader's method of spelling out who, what, where, when and why.

For example, instead of saying, 'There's a meeting this afternoon; it seems that one of the managers is leaving. I bet they'll restructure our department once she's gone. Tim warned us that this might happen. We've got to meet at 2pm and John said that we must all go. I hope this doesn't mean redundancies for any of us' you could start with, 'There's a meeting today at 2pm and John said we must all go.' (The 'headline'.) Then, and only then, elaborate, in the same way that newsreaders do: 'Tim warned us this might happen; that once the manager left, they could restructure our department. I hope it doesn't mean redundancies.'

Being clear and direct has a number of benefits, including the following.

- Other people don't have to second-guess what you really mean.
- Misunderstandings are avoided.
- It enables negotiation to take place.
- It saves time and effort.

Use positive language

What's in a word? Words express ideas, opinions and feelings. Words convey ideas, information and direction. The words you choose reveal a lot about your attitudes and way of seeing the world. Your words have an impact on others.

Two people can express the same idea using different words, and those two messages will take on completely different connotations. Consider the following two sentences:

- 'You need to stop being so anxious.'
- 'You can be more confident.'

In one way, both sentences convey the same message. But, each sentence reveals a different attitude and will have a different effect or create a different impression on the listener.

Most of the time, you probably don't stop to think about whether you frame your remarks in a positive way. This *can* be changed. Be more conscious (and conscientious) about the words you use. Here's how to start.

be more conscious (and conscientious) about the words you use

Listen. You can learn a lot for free just by listening. Listen to people talk on the TV and radio. Listen out for negative words and phrases and try to think of positive alternatives. What difference could your chosen alternative words and phrases make?

Monitor your own speech. If you catch yourself using negative words and phrases, stop yourself, even mid-sentence, and rephrase what you want to say in more positive terms. For example, instead of saying, 'Sorry to hassle you, but I'm wondering if you have made a decision yet?' you could say, 'I'm wondering if you've come to a decision yet?'

Think before you speak. Remember, it's perfectly acceptable to pause and organise your thoughts so that you can phrase them in a positive way.

brilliant tip

It's not just your words that convey a message, it's all of you. Use your body language to support, moderate and reinforce what you say and make sure your mannerisms don't detract from what you're saying.

How you say things

Be aware of the impact tone, pitch, emphasis and pace can have on what you say. Do you, for example, need to slow down? Maybe you've always been a fast speaker. Perhaps you speed up when you're nervous, excited or stressed. Whatever the situation, slowing down will help you to think as you speak (your brain will be able to keep up with your mouth!) so that you can communicate clearly and positively. You'll look and feel in control.

brilliant action: Learn to slow down.

- Practise the pause. Often, the problem is not that you say the words too fast but that you fail to pause at the end of a sentence. People need a few seconds to process what has just been said.
- Use the beat method. Count two beats at the end of every sentence. For example, today is Monday (1, 2), tomorrow will be Tuesday (1, 2).
- Read with rhythm. Practise reading song lyrics or poetry. They have a natural rhythm with built-in pauses. Pause for a comma, and use a longer pause at the end of a verse.
- Write the word 'pause' or 'breathe' on a sticky note. Place the note on your computer or near the phone, to serve as a reminder.
- Ask for feedback. Tell other people you're working on slowing your speech and would like to know if you are speaking slowly enough. Have a friend give you feedback when you start to speed up.

The listener's perspective

So, you're clear about what you want to say and how to say it. Now you need to shift your awareness to the listener. Don't assume that, just because you are saying something clearly and concisely, they will understand what you mean. Consider *their* situation:

- What do they already know or not know about what you are talking about?
- What might their feelings, values or beliefs be?
- Is this a good time and place for them to listen to what you have to say?

Once you have thought about the answers to these questions, you are in a better position to start talking. However, although opening yourself to other people's perspectives makes it easier to get your own message across, it does require imagination and effort on your part. *You* have to do most of the work; other people don't always spell out what they're really feeling. They do, however, give clues. Look for these clues. You have plenty to help you: what the other person says, how they say it, their body language, their actions and how they respond to what you say.

> be prepared to adapt your style when you're talking to someone

Be prepared to adapt your style when you're talking to someone. This doesn't mean that you can't be yourself, but adapting your style makes it easier for other people to listen and to understand what you mean. Look at the example below.

brilliant example

Remember Polly, at the beginning of this chapter? Polly has just been promoted to a supervisory role. Most of the time I talk too much. I don't really think before I speak – I waffle on and use exaggerated gestures which some people find off-putting. I'm worried that I'm going to have difficulty explaining things clearly and concisely to the people in my new role at work.'

▶

Once Polly realised that she was predominantly an Active communicator, she understood how to adapt her style for her new role at work. She retained her friendly, enthusiastic approach, but she practised slowing down her speech, which gave her time to think and explain things more clearly and concisely to her colleagues. Polly also cut small talk to a minimum when speaking with colleagues who wanted to get straight to the point and discuss ideas and plans. She allowed herself more time to chat with others who enjoyed long conversations (whether they were work-related or not).

Finally, Polly toned down her approach with one particular colleague who was even more of an Active communicator than she was. Polly recognised that it's not always the case that you communicate best with those who have the same communication style as yours. Sometimes similar communication styles work well together, but other times they can create a clash!

> If you talk to a man in a language he understands, that goes to his head. If you talk to him in his language, that goes to his heart.
>
> Nelson Mandela

The right time to talk

Effective communication means not just knowing *what* you want to say and *how* to say it. Often, it can mean knowing *when* to say something.

---

brilliant example

Jacqui is a freelance magazine writer. Knowing when to pitch her feature ideas is crucial to getting them accepted. As a general rule, Mondays and Fridays are not the best days to propose new ideas. Editors start the week off with a sizeable workload in front of them and end the week winding everything up.

Annie has learnt that trying to talk about anything with her daughter Harley when she is getting ready to go out in the evening is never going to be successful. Having a chat over dinner or in the car when she's driving Harley to football practice makes better sense. Annie discovered that it wasn't that she and her daughter couldn't communicate; it was just that Annie was choosing the wrong times to converse with her daughter.

Often, there really is a right and wrong time to talk about things. Waiting for the right time makes it more likely that the other person will listen to you and you will effectively get your message across.

This does not mean putting off important conversations. It means recognising that there are times when others are more open and receptive to listening and communicating. And, if you're not sure, just ask, 'Is this a good time to talk about ...?' It's a win–win situation – you'll both benefit.

Invite questions and feedback

Remember, communication is a two-way process. Sometimes you can tell if someone hasn't understood you – not necessarily because they've told you, but from their facial expression. Other times, you can't be certain.

Instead of rattling on, simply ask, 'What do you think?' or 'How does that sound?' This possibility never seems to occur to most people! Ask your partner, colleague, child or whoever you are talking to, 'I'm not sure if I'm being clear, can you tell me what you understood/think I said?'

Asking questions is also useful when you think the other person isn't listening to you. Simply ask them a direct question. Be kind; don't try to catch them out (a favourite trick, I remember, of school teachers). Simply say, 'What do you think?' or 'Would you agree?' Then, when they look at you blankly, just repeat the question and add the topic you were talking about, '*What do you think* the government will do about the housing shortage?' or '*Would you agree* that this restaurant serves the best chips ever?'

brilliant recap

Remember, although there are no hard and fast rules about how to talk to and speak with other people, there is one key principle: make it easy for other people to listen and to understand what you mean. Communicate purposefully; take time to think before you speak. Be more aware of what you say, how you say it and when you say it.

Put it in writing

What positive things can you do with good written communication skills?

Whether you are writing a letter, an email, a report, application form or CV, knowing how to communicate effectively in writing is one of the most important skills you can have in business, professional and formal situations. This is especially true when, it appears, few people possess excellent writing skills; having these skills will help you to stand out and do business more effectively.

People with good writing skills are prized by employers because they are good at communicating with colleagues and other businesses. But not only can well-written letters, emails, reports and other communications help in creating a good impression and prove valuable in increasing business, good writing helps your reader to see you as a real person, and treats them as one too.

Having brilliant writing skills means that you:

- are more likely to get your message across – whether it's to give or ask for information, offer a service or make a complaint
- come across with more authority
- can easily explain complex instructions
- produce natural and fluent text quickly and easily
- put your ideas and thoughts into words

- write in a way that creates a positive impression
- avoid common mistakes to produce accurate, clear writing.

Speaking and writing

How do you communicate effectively in writing? The same way that you do when speaking: clearly and concisely, with confidence and empathy. But, when you communicate in person or on the phone, you have the advantage of being able to clarify or emphasise what you mean and what your intentions are. When you communicate in writing, this doesn't happen, so what you say in writing must express exactly what you mean – there is nothing else to support or clarify your intentions and feelings, requests, demands, information and ideas.

Although you should write in the same way as you speak, your language, words and sentences need to be clearer and more precise. Just as you do in speech, you need to make it easy for the other person (the reader) to understand your meaning. In fact, once you have developed the knack of writing effectively, you'll find that what you say also improves.

Always try to focus on the needs of the reader and make an effort to see things from their perspective. Put yourself in their position and imagine what it would be like for you to be receiving your letter.

Poorly written messages create confusion or fail to achieve their intended purpose. To communicate effectively, to write what you mean and to make sure your reader understands you, follow the guidelines below.

What's your point?

First, think about the main point that you need to make. Put it in a sentence. Then provide details that spell out what, where, who, when and why.

The recipient is more likely to respond in the way you would like them to if you have clearly set out what you want or do not want.

Maybe you want to:

- provide or ask for some information
- invite someone to attend an event
- ask for something to be put right
- ask someone to buy your product or service
- confirm a decision or have a decision confirmed
- agree or refuse to do something.

Whatever it is, for your reader to understand and respond, the purpose of your letter must be clear in your own mind, or your message may get lost.

---

brilliant example

Bad example

Read the email below and see if you can work out what exactly the writer wants.

> Hi. I'm getting in touch about photography. I've been taking photos for many years now (I love doing landscapes and portraits) and I want to get better at it but I'm not sure what sort of equipment or experience I need. I've got Nikon cameras and I've got some experience of Photoshop but are there any other programs you think I would need to know in order to come on the course?
>
> I'm soon going to be made redundant from my job in insurance and was thinking of a complete career change, but I'm not sure how to go about it, so if you have any advice I'd be grateful if you could send it to me!
>
> Cheers, Luke

▶

Did you work out what exactly Luke wanted? It's unlikely!

Here's a good example – this time Luke has explained clearly what he wanted and his reasons for asking.

Good example

> Hello. I am interested in your freelance photography course, but I'm not sure if I have the right level of skills and equipment to join the course.
>
> I have been a keen amateur photographer for several years, but as I have recently been made redundant, I am thinking about making a career change and becoming a freelance photographer.
>
> Please could you tell me what camera equipment and what level of skills and experience I would need in order to be accepted for a place on the course?
>
> Thank you
>
> Luke Harding

Which of the two examples is most likely to get a relevant reply? Or even to get a reply at all? In the first example, it's not clear what the writer wants. The point of the email is lost; the recipient has to work hard and is probably confused about how to answer.

Don't make the reader have to work to understand your letter or email; make it easy to understand and to reply to.

In the second example, the writer has made his purpose clear in the very first sentence. If you don't make your message clear, right from the start, it's easy for the other person either to misunderstand what you said or even to avoid answering.

Vagueness is too common in communication. Rambling, hinting and alluding are indirect ways to say something; they hide the meaning of what you really want to say. The best way to make it easy for others to read and understand is to be clear, direct and concise.

So, ask yourself if your message was clear; did you say what you wanted to say?

the best way to make it easy for others to read and understand is to be clear, direct and concise

Write concisely

Good communication is all about quality, not quantity. We all receive too many communications these days, including letters, emails, adverts and junk mail. The only letters that are read fully are concise. Letters, emails and reports that ramble or are vague will not be read properly.

Concise letters can be understood quickly. So be short. Once you've decided what your main point is, decide what the reader does and, just as importantly, doesn't need to know. Get straight to the point, keep to the point and avoid making the reader work for the point. Save digressions for a face-to-face chat.

First, plan your letter or email by taking a couple of minutes to outline the specific points you want to make, and write a quick first draft.

At this stage, *don't aim for perfection* as this paralyses free thought; just get your ideas on paper, then decide what to remove. There will always be something. It's much easier to edit a letter you've already written than to edit as you write. So, write a first draft, then work on the sequence of your points and on your word choices, style and grammar.

The advantage of communicating in writing is that you can edit your message so that it conveys exactly what you want to say. So, write a first draft to include everything that you want to say. Then rewrite it to ensure that your message is clear. Even professional writers follow this process – they write down their ideas and then they rewrite them until they are happy that what they have written is clear and concise, in a logical order, and conveys what they want to say.

Remember, you want your reader to understand you and respond. Anything that does not help them do that is unnecessary. Avoid repeating anything, other than for emphasis. Remove unnecessary words from every sentence, unnecessary sentences from every paragraph and unnecessary paragraphs entirely.

brilliant example

Clarity comes with simple words and short sentences.

The average sentence contains about 15 to 20 words, sometimes a lot less. This does not mean making every sentence the same length; you can vary your writing by mixing short sentences with longer ones. Most long sentences can be broken up in some way. Aiming for one main idea in a sentence will automatically enable you keep to the average sentence length.

Cut out unnecessary words and phrases. Instead of, I'm quite confident and definitely believe that the event will be a very successful one' write, 'I'm confident the event will succeed.' The second sentence is both shorter and stronger.

Instead of, 'I would be extremely grateful if you could take the time to send me …' write, 'Please could you send me …'

Provide information and get your facts straight

Although less is more when it comes to communicating, don't take being concise too far; there's no need for your written communications to read like a text! Make sure that you include enough detail for your reader to get the message, and that you come across politely, not abruptly.

Start with the relevant information: mention any correspondence the other person has sent you or conversations you've had in relation to what you are writing about. For example, rather than start an email with, 'Hello Kirsty. Attached are details for an idea for a new book' it would be better to write, 'Hello Kirsty. I spoke to you on the phone last week about an idea for a new book. I have attached details.'

Do keep related information together. Don't move on to another point until you've finished the last, and resist the urge to move back and forth between topics or issues, otherwise you'll come across as rambling and confused.

When you have relevant facts, figures, names, dates and examples, make them known. If you are uncertain, say so. If you are merely speculating, say so, so that your reader does not think you know more than you do. Check what you have before you send it, to save the embarrassment of correcting a mistake the reader later brings to your attention.

Spellcheck and reference books such as dictionaries and grammar guides will help you write with correct spelling, punctuation, grammar and formatting. The facts, however, are yours alone. Letters and emails are permanent records, so information, supporting points and facts must be correct.

brilliant tip

Examples make the link faster

If you put forward an idea, always follow it up with an example, to help the reader understand what you mean. 'I think we could do with some fresh perspectives' could be followed with, 'We could, for example, invite the interns to the meetings and ask for their ideas.'

Mind your language

Who will read your message? Is it targeted at senior managers, family, a friend, or a small group of archaeologists? Whoever you're writing to, it's important to use language that's inclusive – that anyone can understand.

In written communication, use a conversational, yet respectful, tone and try to phrase things in a similar way to how you'd actually say them. This doesn't mean using slang or jargon, but there's no need to use stiff, overly formal language. Chances are, you're not a nineteenth-century scholar, so don't write like one. Avoid overly formal terms like 'hereto', 'as per' and 'herewith'.

You should avoid using jargon and slang. Jargon is similar to slang – jargon uses words and phrases which are unique to a small group of people, usually in a particular job or profession, as a kind of shorthand, to express ideas and issues. Most of the time, jargon evolves because the people in a certain job use words and terms which don't mean much to people not working in that profession or particular job.

Such jargon is fine if every reader (or listener) understands it. Typically, though, it is a barrier to clear communication. (It can even be a way of deliberately excluding or confusing others.)

For example, a reference to a 'CPN' (community psychiatric nurse) or to 'meds' (medication) or 'Obs' (Obstetrics) will make sense to a healthcare professional, but if the reader doesn't come from a health background the writer needs to explain the reference. Always put yourself in the place of the reader and, if there's any doubt, explain the jargon in plain English.

Jargon is fine when it represents a concise way of saying something to people who can make sense of it, but not when it's a substitute for easily understood words.

Using gobbledygook, such as 'comfort break' rather than 'break', 'blue-sky thinking' for 'clear thinking', 'citizen empowerment' for 'people power' or 'slippage' rather than 'delay', is not good English either. It's unnatural and contrived, and is difficult to understand.

Finally, when thinking about the words and phrases you include, avoid using condescending or patronising language. Cut out phrases such as, 'As I'm sure you're aware …' or 'Of course, you must know …' Always assume that the recipient is intelligent but uninformed, not stupid.

brilliant impact

Since 1979, the Plain English Campaign (**www.plainenglish.co.uk**) has been campaigning against gobbledygook, jargon and misleading public information.

They define plain English as 'writing that the intended audience can read, understand and act upon the first time they read it'. They also describe plain English as 'a message, written with the reader in mind and with the right tone of voice, that is clear and concise'.

The Plain English Campaign has helped many government departments and other official organisations make sure their public information is as clear as possible.

Everybody understands words that apply to everyday life, so use everyday words and your reader will understand you.

brilliant example

Three grammar rules it's (sometimes) OK to break

Grammar is one of those things that, when not used correctly, can make you look very stupid very quickly.

That's not to say that all grammar rules are written in stone; there are times when they may be broken. But only sometimes. Do not think you can break the rules all the time – only when they make a sentence flow better.

Here are three grammar rules it's OK to break, sometimes.

You can start a sentence with a conjunction. Somebody, somewhere, once decided that you shouldn't begin sentences with conjunctions. But you can start a sentence with *and, but, however, so* or *because.* ('Because she didn't like spinach, he never bought it.') It's a great way to grab attention and emphasise a point. (Watch the film *Finding Forrester.* Sean Connery and Rob Brown's characters have an entire conversation about it, and deliberately start their sentences with the offending words in order to make their points.)

You can split infinitives. So you can say '*to boldly go*'. 'To go boldly where no man has gone before' just doesn't have the same ring to it as 'to boldly go'. In other words, if you know what you are doing and you try the phrase the correct way and it sounds daft then it is fine to split infinitives.

You can end a sentence with a preposition. In English, the most used prepositions are *of, to, in, for, with* and *on.* A preposition indicates a relation between things mentioned in a sentence. For example, 'Who did you go with?' sounds less formal and more conversational than 'With whom did you go?'

Remember, this does not mean you should break rules all the time – only when they make a sentence flow better. As my friend Sue, an English teacher, explains, 'Breaking the rules should never be an excuse for sloppy English – but a deliberate decision because it sounds better than using it the "correct" way or because it is the effect you want to achieve.'

Write actively

Aim to use an active voice; you will come across as clear and professional rather than stuffy and formal. Having an active voice means using active verbs.

What's the difference between active and passive verbs? This is best explained by understanding how a sentence fits together. There are three main parts to almost every sentence:

- a subject (the person, group or thing doing the action)
- a verb (the action itself)
- an object (the person, group or thing that the action is done to).

To give an example, in the sentence 'Amy peeled an apple' the subject is Amy (she is doing the peeling); the verb is peeling; and the object is the apple (it is being peeled).

There could be lots of other words as well. For example, 'Amy, the girl I sat next to in class, always peeled her apple before she ate it.' But the subject, verb and object are still there.

With an active verb, the three parts appear in a specific order: subject then verb then object. For example, 'Amy (subject) peeled (verb) the apple (object).' 'Peeled' is an active verb here. In an active sentence, the subject is doing the action. In active sentences, people do things; they act and interact. An active sentence is simple and brief, describing who acts and how.

With a passive verb, the order is reversed – object then verb then subject. 'The apple (object) was peeled (verb) by Amy (subject).' 'Peeled' is a passive verb here. The sentence says what is being peeled before it says who is doing the peeling. The words *was* and *by* have appeared and the sentence doesn't flow as well.

In 'The solicitor represented her client' the verb 'represented' is in the active voice, but in the sentence 'The client was represented by his solicitor' the phrase 'was represented by' is passive.

Overleaf are some more examples of how to turn a passive verb into an active verb.

The computer was fixed by Anne. (passive)
Anne fixed the computer. (active)

This matter will be considered by us shortly. (passive)
We will consider this matter shortly. (active)

The source of your problem has been determined. (passive)
We have found what caused your problem. (active)

an active voice is more lively and to the point. It's also easier to write and understand

Passive verbs make sentences long-winded. Also, the passive voice is vague and evasive, making your reader uncertain who is doing what. An active voice is more lively and to the point. It's also easier to write and understand.

Good uses of the passive voice

There are times, though, when it is appropriate to use a passive voice. For example, to avoid taking the blame – 'a mistake was made' (passive) rather than 'I made a mistake' (active). Or, to avoid appearing too harsh – 'this bill has not been paid' (passive) is less abrupt and accusatory than 'you have not paid this bill' (active).

This difference between active and passive verbs is not easy to grasp: just aim to sound natural and conversational and, nine times out of ten, you'll get it right.

Be positive and upbeat

Do you use negative language and phrases? If so, aim to use constructive, positive language. Words and phrases that affect your reader positively will help them to respond positively, whereas writing with a negative attitude is likely to generate resistance and defensiveness.

brilliant example

Positive language

It is not uncommon for organisations to send letters that are phrased in negative language.

> We are writing to inform you that we are unable to process your application, since you have neglected to provide all the relevant information.
>
> You will need to complete ALL sections of the attached form and return it to this office within the next 14 days, to avoid unnecessary delay to your registration.

This is overly formal. It's negative; it implies wrongdoing on the part of the recipient instead of suggesting what they can do to fix the problem. All the information is included, but it comes across as abrupt and formal.

Here is the message again. It contains the same information, but is written in a more positive way – upbeat and helpful.

> Thank you for your application. To complete registration, we need some additional information. If you return the attached form, with the highlighted areas filled in, we will be able to send you your registration papers within two weeks. We look forward to hearing from you.

Positive phrasing and language:

- tells the reader what *can* be done rather than what *can't* be done
- suggests alternatives and choices
- sounds helpful and encouraging rather than formal and abrupt
- emphasises positive actions and positive consequences.

Avoid using negative words and phrases. Here are a few common examples.

Words that imply fault:

you didn't

you overlooked

you neglected to include

you failed to provide

Phrases that suggest the person is not being honest:

you claim that …

you say that …

Domineering phrases:

you should …

you ought to …

you must …

I must insist …

I expect you to …

unless you …

Uncooperative words:

can't

won't

shan't

mustn't

haven't

Cut out negative words and phrases and replace them with more positive messages. For example:

'If you can send me … I can complete the process for you.'

'One possibility is for you to …'

'I can help you to … if you can send me …'

'I would like …'

'If you would like …'

'I will be able to …'

Even if the message you have to convey is negative, you can soften its impact – you don't have to be heartless. Common techniques of doing this include using the positive form of a sentence. For example, instead of saying, 'You can't work on Project A' a more positive message is, 'You can work on Project B.'

Instead of, 'I won't know until tomorrow' write, 'I will know tomorrow.'

Your writing should encourage your reader to respond positively. To maintain positivity, suggest possible actions rather than lengthy explanations, excuses and too many apologies. People don't want to know what you can't do – they want to know what you *can* do. Focus on that.

remember, your aim is to be understood and to get your reader to respond in the way you hope

Remember, your aim is to be understood and to get your reader to respond in the way you hope. Negative language is, at best, demotivating and uninspiring; at its worst it can create the wrong response: hostility or confrontation. Positive writing is constructive – it encourages interest, a will to understand and a positive response.

Don't avoid negative words completely, however. Negative language has its place. For example, 'Drive sober' just doesn't have the same impact as 'Don't drive drunk'.

Think about what you write and eventually your thoughts, speech and behaviour will become as positive as your language.

brilliant action: Practise a more positive vocabulary

Go back over a few recent letters and emails you have written. Read through each one highlighting sentences that come across as negative. See if you can reword negative statements and phrases to make them positive.

In future, review what you write before you send it. Written communications are the simplest place to start building a more positive vocabulary because, unlike speech, you can edit what you say. Once you have developed the knack of writing positively, it will be easier to change your spoken language to be more positive.

brilliant tip

Collect examples of good writing – what do you like and dislike? Don't just look at the content (the words and phrases that are used). Look at the layout too. What makes something easy to read?

Structure and layout

The right use of words and sentences, grammar and language helps to make it easy for the reader to read, and for you to get your message across. How your writing is structured can either help or hinder. Here are some ideas for structure and layout that will help you.

- Use one idea for each paragraph. Don't confuse the reader, or yourself, by including more than one point in a paragraph. Prioritise the most important points and put them in a logical order.

- Use plenty of white space to organise text. Avoid long, dense paragraphs – they can be very intimidating and discourage the reader. Remember, you want to make it easy for the reader.
- Use headings to divide the text into logical sections. This makes it easier for the reader to quickly scan and get an overview. It also helps the reader to reread and easily find the relevant information.
- Use bullet points to identify key points. This is another good way to separate out ideas and information. However, avoid bullet points when you need to deal with a sensitive issue. Bullets communicate efficiency rather than warmth.

avoid bullet points when you need to deal with a sensitive issue. Bullets communicate efficiency rather than warmth

- Use bold type, italics or underlining for emphasis (but don't overdo it).

Be more aware of everything you read; look out for good and bad examples of writing. Practise writing clearly and concisely. Use positive, inclusive language. You will soon find it quick and easy to produce natural, accurate, fluent text that will get your message across and be easy for your reader to understand. And if your reader can understand and respond, then your writing has been successful!

PART 2

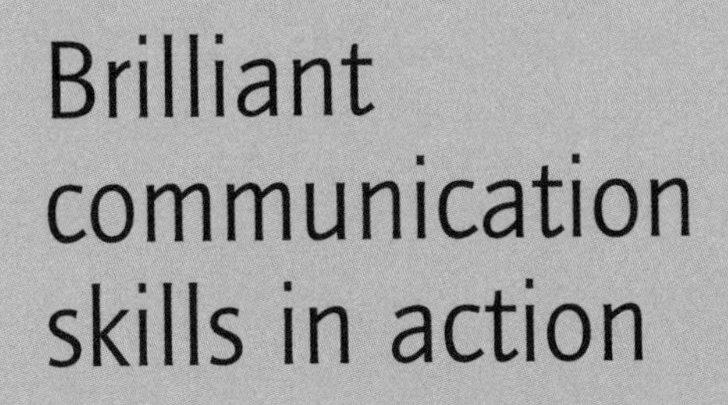

Brilliant communication skills in action

Putting it into practice

What to say to motivate, encourage, persuade and influence others

> Our chief want is someone who will inspire us to be what we know we could be.
>
> Ralph Waldo Emerson

In Part 1 you will have learnt what makes for brilliant communication skills and how to develop them. You will also know that communication is rarely straightforward. So often it's dependent on context; the circumstances and situation in which communication is happening. In Part 2 you will find specific situations and see how those communication skills work in practice.

Knowing how to motivate other people means knowing how to guide them towards specific goals. Whether those goals are initiated by you or by the person themselves, your aim is to give them the confidence and belief to achieve those goals.

You can get people to work on and achieve goals by making demands or being manipulative, but that's forcing people to achieve things in a way that works for *you* and probably only serves *your* interests. You might succeed in getting things done, but you won't succeed in winning support.

Whether you are hoping to motivate your teenage son to clean his room or persuade a group of colleagues to follow a particular course of action, there are several key communication principles you must follow.

1. To begin with, ask yourself what you already know about the other person that might help you to engage them. What's their communication style? Do they, for example, respond well to conversations that take place while you are both involved in an informal activity?
2. Be clear about your message. Make sure you know what exactly you want to persuade the other person to do; what you want to get from the conversation, email, phone call or speech. Simplify the message and don't ramble on. People who successfully motivate others make it easy for them to understand.
3. Listen. Communicating effectively involves both being understood and being understanding. Listen not just to what is being said, but also to what is *not* being said. That is what good motivators and influencers do. Tell the other person what you want and then listen to their response. Be a good listener and take their point of view into consideration. People are far more willing to cooperate if they feel acknowledged, understood and appreciated.
4. Ask questions. Find out what the other person wants. What incentives do they need? What are their concerns? Acknowledge and address those needs and concerns. Persuasion requires and involves emotion, but don't let feelings take over. Discuss ways to overcome any barriers.
5. Once you've addressed needs and concerns, accentuate the positive. State the benefits, explain what's in it for

them, and deliver it as the most attractive possibility. For example, 'Put your dirty clothes in the laundry basket every night' is likely to mean a lot less than the message, 'If you put your dirty clothes in the laundry basket each night, they'll be washed and dry and ready for you to wear the next day.'

brilliant example

Lee is the manager of an adult education organisation. He had to persuade all the freelance tutors to join a national professional association for teachers, trainers and tutors. The problem was that the tutors had to pay the annual membership fee themselves.

From formal appraisals and informal conversations, Lee knew that apart from being well paid and securing plenty of work, what was important to the tutors was ongoing training and professional status.

Lee listened to their concerns and reflected back what he had understood. He explained that although he understood the tutors resented having to pay a fee out of their own pocket, he felt they would get a good return for their money. Lee pointed out the benefits: if they agreed to join the professional association, tutors would have access to a wide range of training, run exclusively by the professional association for tutors' professional development. This would enable them to keep up to date with their vocational or subject specialism, as well as the latest practice in excellent teaching and training methods.

Lee said that the association was working hard on the tutors' behalf to influence national decision-making and policy about a pay increase. He also pointed out that other adult education organisations required membership of this association, so if the tutors signed up it would increase their employment opportunities.

The tutors were convinced – they all took out the membership!

Giving advice and information

Whether it's asked for or not, advice can be either hurtful or helpful; it can disrupt relationships or it can have a positive effect. It can provide new ideas and information, and help the other person identify their options, make a decision, move forward and take control.

> advice can provide new ideas and information, and help the other person identify their options

To make sure your advice is well received, bear in mind the following.

1 Listen carefully. Although you may have some insight or have experienced a similar problem, you need to listen carefully to the other person first. Listen and learn as much as possible about their situation. If you need clarification, ask.

   Being an active listener will not only help you give good advice, it will also make it more likely that the person will take your advice. Why? Because you have listened and checked your understanding, it will be *relevant* advice. But if you don't really feel qualified or knowledgeable enough about something to give advice on it, say so.

2 Having listened carefully, and if you feel you can help, make sure that the other person actually wants your advice. They may just want to offload and for you to listen. Be a supportive listener, and wait for your cue to respond.

   Check with the other person what exactly they want advice about – what is their goal? Most importantly, ask them, 'What do you want to happen?' so that you can tailor your advice directly to the person's stated goal or outcome.

   Ask, 'Do you want some ideas to improve the situation?' or 'Can I suggest something?' or 'Can I give you my opinion/advice?'

3. Keep your advice short and to the point. Tailor the advice directly to the other person's stated goal or outcome. Even if there appears to be a correct answer to a problem, think what other possible courses of action there might be and, together with the other person, consider the pros and cons of each. Help them to come to their own conclusions and decisions. When you've talked for a few minutes, bring it back to them: 'What are your thoughts about that?'

4. Avoid judging. Think back to the last time you gave someone advice. Did you express concern and understanding or were you frustrated, angry or worried about their situation? When you feel like this, you may think you are giving good advice, but you are probably coming across as negative or critical.

   Use positive language. Rather than start your advice with, 'Why didn't you …?' or 'You should have …' accept what's done is done, and focus on what they can do next. Try something like, 'How about …?' or 'It might help to consider …'

---

brilliant example

If you've been in a similar situation, think about what you learnt. Then, rather than say, 'That happened to me, you should do what I did', simply share your experience, by asking, 'Would you like to hear how I managed when that happened to me?'

For example, imagine someone you knew confided in you about a health problem they had that you had also experienced. The other person's feelings and emotions about the situation might be different from yours, but by sharing your experience they might be able to pick up some insights rather than feel they have been *told* what to do.

Use the power of example rather than the force of knowledge!

---

5 Know when to let go. Even if the other person asks for your advice, they won't necessarily take it. Their feelings and emotions about the situation might be different from yours, so you can never be sure that your advice is right for them. Don't be surprised if a person rejects your good advice and decides to follow their own (potentially disastrous) course of action.

Giving compliments and positive feedback

Giving compliments and praise, expressing appreciation and gratitude are all forms of positive feedback. Not only can this brighten someone's day, but positive feedback can also encourage and motivate the other person to keep up the good work.

When you praise an employee or colleague, show appreciation to a friend or family member, or say thanks to a company or individual for good service, your remarks will let the other person know that their actions have been noticed. This can develop empathy and rapport because it shows that you have made an effort too; you have bothered to acknowledge their actions. Putting it in writing shows even more effort on your part while also giving the person a permanent reminder of the praise.

> a genuine sentiment phrased awkwardly is better than no appreciation being showed at all

Too often, though, you can worry about getting the wording right and so end up not saying anything at all. Just remind yourself that a genuine sentiment phrased a bit awkwardly is better than no appreciation being shown at all.

To help you, follow these steps.

1 *Start with the reason why you are complimenting or praising the reader.* Be specific. Sometimes the most memorable

compliments are the most specific ones, because it shows that you were paying attention. For example, 'I want to say thank you for helping to look after my mother during her stay here' or 'I want to congratulate you on your success in getting the funding for our next project.'

2. *Acknowledge personal qualities or special efforts*. For example, 'Your concern and patience with Mum and your sensitivity to her needs have been much appreciated by me and my family' or 'Your expertise in this area and your willingness to put in the extra time was brilliant.'
3. *Explain what a difference their efforts have made*. People feel good if they know that they have made a difference. So, if what they have done has had a positive effect on you, tell them! 'Your concern helped me feel better.' 'Your support has reassured me.'

   When you tell the other person that they have made a positive difference, they can then feel good about themselves and encouraged because of the impact their actions had on you. And there is no way they can deny your compliment without sounding silly: 'No, it didn't!'

   Be aware, though, that for some people and in some cultures, it is polite to deny compliments and impolite to accept them. Sometimes a person may refuse a compliment for this reason. Don't press the point – the appropriate response from you is a smile and a simple nod of your head.
4. *Express appreciation*. 'Thank you.' 'I really appreciate it'.

   Finally, if you're giving a compliment or expressing appreciation in person, don't forget: your body language, tone of voice and facial expressions will show that your compliment or appreciation is genuine.

Accepting compliments and praise

The best way to accept a compliment is like a gift; just say 'thank you'. If you want to add something, make sure it is positive.

- 'How nice, thank you.'
- 'That's the best thing anyone has said to me today (this week/in a long time). Thanks!'
- 'Thank you for telling me.'
- 'Thank you. I really appreciate hearing that.'
- 'Thank you. I'm pleased too.'

If you receive a compliment on a job well done and you didn't do it alone, make sure you acknowledge those who helped. 'Daryl and Ruth were such a big help. I couldn't have done it without them.'

How to make small talk

Small talk can be a big challenge! Small talk doesn't come easily to all of us; often, the fear of coming across as fake, dull or stupid makes it difficult to screw up your courage and initiate small talk, let alone maintain a conversation.

But being able to make small talk can help you to create all kinds of connections. It can open doors to all sorts of people and possibilities. Small talk can lead to big talk, which can lead to big people and big opportunities!

the golden rule is to be, or at least appear, friendly and approachable

Remember, good communication skills require confidence and empathy. The golden rule is to be, or at least appear, friendly and approachable. Your aim is not to impress, it's to show that you are a relaxed and sociable person who is open to exchanging a few pleasantries.

Be positive. Don't start off with a moan or a complaint; say something positive. Smile, make eye contact and keep your body and gestures open and relaxed. Don't stand too close to people. Do not stand too far away, otherwise they will think you are afraid of them or stand-offish.

Making the first move

Initiating conversation is a bold step; fear of rejection is the main reason why people don't make the first move. If you're at a party, convention or any other social gathering, choose a person who seems approachable – someone standing by themselves is a good bet – then just look them in the eye, smile and say, 'Hi, I'm [give name]. What's your name?'

brilliant tip

Do you forget names easily? It will help you to remember a person's name later in a conversation if you repeat their name when you are introduced to them: 'Nice to meet you, Hannah.' If you later find that you've forgotten their name, admit it. 'Gosh, I'm sorry, can you remind me of your name?' Once they've told you, don't make a big deal of it, simply repeat their name and move on with the conversation. 'Hannah. Thanks. Well, as I was saying ...' It won't be nearly as tough as you think, and remember: the longer you leave it, the more embarrassing it gets to admit you've forgotten.

Asking good questions

One of the most important qualities for engaging another person in a conversation is your curiosity and interest. Your interest will be apparent in the way you ask questions, a crucial communication skill (see Chapter 3).

You don't need to have the perfect opening questions (even if you start with the usual questions, 'What do you do?' 'How do you know Rebecca?' 'Have you been here before?') but you do need to be interested in the answers.

Ask open questions that require more than a 'yes' or 'no' answer, and give the other person more of an opportunity to talk about themselves, their opinions or experiences.

Listening well

Supposing you started by asking, 'What do you think about all this rain?' Listen for something in the other person's reply that might suggest a direction for the conversation to take. For example:

You: 'Can you believe all this rain we've been having?'

Other person: 'Oh, I don't mind the rain. It's good for the garden.'

You: 'Do you like gardening?

Or:

Other person: 'I'm just glad I don't live by the river any more.'

You: 'Where was that? Did you get flooded?'

Follow up on what the other person says. If they say they're feeling great, ask why. If they say they're exhausted, don't just reply, 'Oh dear, are you?' Follow up on it; ask why. Ask questions and, more importantly, listen and respond to the answers.

---

brilliant example

Cultural differences

Why is the weather a common topic? Because it's changeable and unpredictable. Also, it's a safe, neutral subject to start with. Talking about the weather lets you and the other person take small conversational steps until you find common ground and can move on to other subjects.

Questions about family are usual in some cultures, although it may be inappropriate for one man to ask about another man's wife. In cultures that are status-oriented, such as China and Japan, small talk between new acquaintances may include several questions that help establish each other's social status.

In some cultures, conversation moves quickly to the point; in other cultures, people talk about other things long enough to make a connection and establish a rapport with the other person. And, in some cultures, meetings usually begin and finish with small talk but have a clear transition to the main purpose of the meeting. Other cultures might expect small talk not to happen until the end of a meeting.

Listen out loud. Show that you're listening by making eye contact and nodding, and by using verbal cues ('Uh-huh', 'Really?!', '*He didn't!*') – all let the person know you're listening and you're interested. The more positive feedback you give, the more others will enjoy talking to you.

Say something. Although you might be genuinely interested in what the other person has to say, don't keep firing out questions, or the other person will feel that they're being interrogated! 'How do you know Sarah?', 'What do you do?', 'Where are you from?', 'Where do you live?', 'Where do you work?' That's too many questions in one go; think how you would feel if you were asked that many questions!

don't keep firing out questions, or the other person will feel that they're being interrogated

But maybe you hesitate to talk about yourself. You may not be comfortable with opening up or you're concerned you might come across as self-centred; you find it easier to turn the focus on the other person. Don't forget that communication is about an *exchange* of ideas, information and feelings – it's two-way, so make a contribution yourself.

Don't worry about coming up with clever conversation starters. Just about anything you find curious or interesting can start a conversation and keep things rolling.

Try commenting on something that both you and the other person are experiencing; your environment and what's around you. For example, instead of saying, 'Don't you just love this restaurant?' which can be answered with a conversation-killing 'yes', say, 'I really love this restaurant.' Even if they're just being polite, it's likely they'll ask you why, which opens up another opportunity for conversation.

By drawing on shared experiences to create conversation-starting questions or statements, it's more likely the other person will have an opinion on the matter to share in response.

Say something about what you're interested in and then ask for their opinion. You might, for example, say something about a book you are reading or a blog or website you've found interesting. What about a film you have recently seen, something funny, useful or interesting you heard on the radio or TV? Or an interesting theory you recently heard? Did you lose your keys or find £10? Maybe you ate at a new restaurant recently, or heard some great new music. Tell the other person, then ask if they've had a similar experience.

Your work, travel, news stories, celebrity gossip: everything is a source of information, ideas and experiences that can be discussed.

---

brilliant example

My friend Zara was driving to a fundraising dinner when she heard a discussion on the car radio that asked people to choose between two superpowers: would they rather have wings so that they could fly, or own a cloak that, when worn, would make them invisible?

'I mentioned it when things had gone a bit quiet,' Zara recalled, 'and just as they had in the radio programme, men preferred the wings and women the cloak. It really opened up the conversation; it led on to several other topics.'

Go off on a tangent. Small talk does not have to end at the topic it started at. Pick up clues from questions and answers and use them to move the conversation on.

Often, awkward silences appear in conversations because you are worried about saying the right thing. *Learn to stop looking for the perfect conversation* and instead just say whatever comes to mind. If you feel like talking about the pizza you had for breakfast, do that. If you follow it up with a question, 'What's the weirdest thing you've had for breakfast?' you've opened up the conversation. You'll discover that it's fine to let go and talk about whatever you feel like.

It's not a competition. You do not have to try to top the other person's stories with fascinating tales of your own. The goal isn't to prove to the other person how exciting and interesting you are. The goal of small talk is simply to make a connection and see where it takes you both. Smile, be interested, ask questions, take a genuine interest in other people's lives and say something about your own.

smile, be interested, ask questions, take a genuine interest in other people's lives and say something about your own

Once upon a time, it was considered impolite to bring up religion or politics, money or health (yours and theirs). Now, pretty much anything is up for discussion, but do look for any clues (particularly non-verbal signs) that the other person may be feeling uncomfortable.

Also, *know when to stop* and pull out. If the conversation feels like climbing a hill of sand then it may be time to move on or let silence take over. You can't connect with everyone, and some conversations simply refuse to take life! Making small talk involves knowing when it's time to let go.

Either way, end the conversation with something nice. For example, 'It was nice talking to you' or 'Have a good evening.'

brilliant example

From silence to small talk to more meaningful conversation

Last week I was given a lift home from Bob, the friend of a friend. We had made a little bit of small talk and then there was a quiet period as we drove through the countryside. We passed a cyclist. I live in Brighton and I remarked that the London to Brighton bike ride was happening that day. Bob told me he used to live in Crawley, which is one of the towns on the event's route. I was born and brought up in Crawley and this one remark, triggered by remarking on a cyclist, was the beginning of an interesting conversation.

brilliant tip

If you're still terrified by the idea of talking to strangers, challenge yourself to talk to one stranger a day, every day, for 30 days. At a shop, café, cinema, theatre, with anyone who works with the public – because they're used to people making small talk. The point of this exercise is to get you used to talking to people you don't know and form the habit of starting a conversation.

If you do this enough, you'll eventually get comfortable being yourself around people. While you should expect your first time to feel strange and scary, look at the big picture. Do you really think you'll still be just as shy once you've done this ten times? Twenty times?

Networking

Networking involves exchanging information, contacts and experience. Whether it's for social or professional reasons, effective communication skills form the heart of networking.

Today, technology has added a new dimension to networking. With only a person's name, you can Google them, look at their LinkedIn profile, their Facebook information and, if they tweet, their Twitter stream. But online social networking doesn't detract from the importance of networking face to face with other people.

Yes, networking takes time and effort, but there are other people out there who want to build genuine, mutually beneficial relationships.

Who to network with?

There are two opportunities: you can network with people who share the same ideas and interests (professional or social) as you; or you can expand your horizons by networking with new people about new ideas and interests.

If you're not feeling confident and think you need practice with networking, start on comfortable ground; do something that interests you. Find an organised club; look through websites and social networking sites, to find other people who share some of your interests and goals and who you can meet up with. Love gardening? Find a gardening club or talks and lectures run by, for example, the Royal Horticultural Society or the National Trust.

Volunteer for a cause that you feel strongly about. Put yourself out there. When you are busy doing something that is fun and meaningful, your networking skills and ability to form connections will start to feel natural.

put yourself out there

Also, people who are shy or introverted tend to be a lot more open and talkative when they're doing or talking about something they're interested in or passionate about. If you find people who are just as keen on board games, hiking or singing as you are, then you'll have a much easier time establishing connections.

Next, expand your comfort zone. Look for opportunities to meet new people and enjoy new experiences. Successful networking, whether for business or pleasure, requires widening your horizons. Talk to people who work in a different department from you, or take up a completely new hobby or interest – one that involves a high level of social contact – a book or film club, wine tasting society or potholing club.

Be prepared to meet people you don't like; when you're networking, you're going to have to sift through the people you don't want to know to meet the people you do want to know.

How to network

Find out who knows whom. When you're talking to people, find out what they do for a living and for fun, as well as what their partner, sons and daughters and close friends do for work and recreation.

brilliant example

You meet Julia at your book club. It comes up in conversation that her sister is an expert horsewoman. A few weeks later, your nephew mentions he's always wanted to learn to ride a horse. You phone Julia and ask if her sister would be able to give your nephew a private lesson as a birthday gift from you. Julia's sister agrees; your nephew is thrilled. A while later, Julia needs some help marketing her new business. Your nephew's girlfriend is a freelance social media marketing consultant; Julia employs her. Your nephew's girlfriend is very grateful and, in return, introduces you to a photographer who takes some excellent photos for your new website. And on it goes!

Find the extroverts. As you network, you'll find that some people are very good at it; they seem to know everyone! Get to know these people first because they can introduce you to others who share or widen your interests or goals.

Be nice. Since you're looking to create mutually beneficial relationships, make the first move and do something to make a contribution. You can offer information, advice and compliments. You can also invite people to a meal or occasion at your home and introduce new people to each other. As long as you're honest and genuine, you're establishing good relations with people and opening future opportunities.

Reach out and ask. The next time you need something, cast a wide net and see what happens. Remember to make it easy for others to understand what you need. Then, make a few phone calls or send out an email stating your situation. Don't apologise. Don't say 'sorry to bother you'. You have nothing to be sorry about; you're simply asking if anyone has the information or resources to help you.

Keep in touch. Maintain your network. If, for example, you read or hear something that might be of interest to someone you know, send it their way. If it's something bad, send a message of commiseration or sympathy. If you hear about something good that's happened to them, send a message of congratulations.

Giving presentations

If the polls and surveys are to be believed, public speaking is one of the things that people fear the most. Whether it's giving many people find public speaking terrifying

a speech at a social occasion or at a business presentation, many people find public speaking terrifying.

Why is this? There are a number of reasons, all related to negative thinking. Thinking such as 'Something will go wrong', 'I'll forget my words', 'I'll talk too quickly', 'People will get bored', 'Everyone will think I'm stupid if I can't answer their questions' and 'If it's not a complete success it will be a complete failure.'

Fixating on these negative ways of thinking evokes memories, images and feelings in your mind that create anxiety and can undermine your confidence. And yet, confidence is exactly what you need: you need to believe that you *can* do it.

Well, the good news is that with a reasonable amount of preparation, you *can* do it!

As well as confidence, a good presentation requires you to have empathy. Why? Because if you put yourself in the place of the audience, you will have a good understanding of what your presentation needs to have – in terms of structure, content and delivery.

brilliant example

Being an empathic presenter

Think about the last presentation you attended, whether it was at a conference, at a meeting, or a speech at a social occasion. What made it good? Was it clear and concise? Was it easy to follow – did you know what the presenter's main points were? Was it interesting? Entertaining? Did you learn anything?

What made it bad? Did the presenter ramble and waffle? Was it difficult to follow – did you know what the main points were? Was the presenter dull and boring? Patronising even?

Learn from past presentations you have attended.

Just like any other communication, the quality of a presentation rests equally on what you say (content) as how you say it (performance). This means making sure that your message is clear and that you have made it easy for your audience to understand.

There is no substitute for preparation.

Preparation

Decide, first, what the purpose and main message of your talk is and, second, the main points you want to make. Make a list of these two things as your starting point.

Good presentations then follow this formula: tell the audience what you are going to tell them, then tell them, and then tell them what you have just told them.

What this means is that you start with an introduction, telling your audience what your main points will be. Then you talk them through your main points, giving evidence and examples to illustrate each point. Finally, you summarise what you've just talked about.

Write out the presentation in rough form. Review it. You will find things that are irrelevant or unnecessary. Delete them. Check it flows smoothly and logically from one point to the next. If there is something you cannot easily express, possibly because you don't clearly understand a specific issue, consider leaving it unsaid.

Next, prepare cue cards with the key points. Use bullet points, keywords and phrases as prompts for all the important points you want to make. This way, you can glance down at the cue cards to remind you of your next point, and then elaborate with free speech. This might seem nerve-racking at first but if you rehearse well beforehand, a quick look at a bullet point on your prompt card as you talk should be enough to remind you.

Never read from a script. This might help avoid hesitations and rambling deviations, but it's difficult to communicate well with your audience if your head is stuck in a piece of paper!

Don't forget to number your cards in case you drop them. But, if you are using PowerPoint slides, then these will serve as your prompts and as a guide for the audience.

Rehearse your presentation – to yourself at first and then in front of a friend or colleague. Ask for honest feedback – what is good and what can be improved?

Anticipate the questions you're likely to be asked and be ready for them with concise and credible answers. Adapt them for the particular audience you're addressing.

Making the presentation

Keep to the time allowed. If you can, keep it short. It's better to under-run than over-run. Stick to the plan for the presentation, don't be tempted to digress – you will eat up time and could fail to make all your points.

At the end of your presentation ask if there are any questions. Listen carefully. If a question is complicated, rephrase it to simplify it without changing the meaning. If it's angry, recast it in neutral language. For example, 'So you want to know if, in the light of recent reports in the press, the company will reconsider its position on redundancies?'

As you answer a question, begin by looking at the questioner, then look at the rest of the audience to show that the answer is meant for everyone. You can discourage persistent questioners from asking repetitive follow-up questions — and give someone else a chance to ask their question — by looking at a different area of the room as you finish your answer, and invite a question coming from there.

Delivery

Speak clearly. Don't shout or whisper – judge the acoustics of the room and, if you're unsure, ask if everyone can hear you.

Don't rush: try to speak fairly slowly and remember to pause after each key point. Pausing at key points also has the effect of emphasising the importance of a particular point you are making. Listen to Barack Obama – he does this very effectively.

Use your hands to emphasise points but don't indulge in too much hand-waving or moving around. Ask the person you are rehearsing with to give you feedback on your body language.

Avoid moving about too much. Pacing up and down can unnerve the audience, although some movement helps to keep the audience engaged.

brilliant tip

Watch how stand-up comedians make use of their body language and stage space. Comedians like Michael McIntyre and Billy Connolly move around a lot. Anyone old enough to remember Dave Allen will recall that he delivered his whole routine perched on a stool.

Which style do you prefer? Which style feels more comfortable to you?

Look at the audience as much as possible, but don't fix your sights on one person – it can be intimidating. Pitch your presentation towards the back of the audience, especially in larger rooms.

Face your audience when talking, not the whiteboard, flipchart or display screen behind you. (Which is what I did recently – despite having been told that one person in the audience relied on lip-reading. The situation was resolved when it was suggested that a colleague could write on the board while I concentrated on the talking.)

keep an eye on the *audience's* body language. Know when to move on and when to stop

Keep an eye on the *audience's* body language. Know when to move on and when to stop.

PowerPoint slides

PowerPoint slides can help provide a clear and concise structure through which to express your ideas and information. However, your audience doesn't need another boring PowerPoint presentation. Even though you might be nervous about public speaking, don't make the mistake of just writing your text on the screen and then reading it out. This is not communicating with your audience. If it's all on PowerPoint, people might wonder why you don't just sit back and watch like the rest of the audience!

Your presentation should be thought out *before* you think about the slides. That way, you avoid letting the slide show dictate the content and structure of your presentation.

Always keep in mind that the slide show is there to *support* your presentation. Rather than make your talk unnecessary, it should need your narration to make complete sense. So bear in mind the following.

- Keep text to a minimum. Too much and you risk making the slides difficult to read; your audience will struggle to read and listen to you at the same time.
- Aim to limit words to a maximum of 35 per slide and bullet points to 5 per slide. Use a reasonable font size and a typeface that will enlarge well.
- Where possible, express your ideas and points with pictures and images. They can create instant impact and can be easier to remember.

Finally, to keep the audience engaged, if you are not actively talking about it, switch off the display, or replace the slide with some form of 'wallpaper', an image or logo.

brilliant dos and don'ts

Do

- ✔ Avoid being too ambitious – don't try to cram too much into the time you have available.
- ✔ Keep it simple – remember a basic rule of communication: plan to get a *few* points across *fully*.
- ✔ Write out notes to refer to during your talk – use bullet points and keywords as prompts.
- ✔ Start with an introduction – begin with a brief summary of what your presentation is about.
- ✔ Use examples to illustrate and emphasise your ideas and main points.
- ✔ Ask questions. It helps to keep the audience engaged and focused.
- ✔ Summarise again at the end of your talk.

Don't

- ✘ Write out what you have to say in long-hand and then read it out word for word.
- ✘ Try to let your audience know how much work you have put into the presentation or how clever you are. People are only interested in understanding what you have to say and learning from it.
- ✘ Rely on PowerPoint to do the presentation for you.
- ✘ Give out handouts during your talk, people *will* read them instead of listening to you!

If you're well prepared, you should feel positive and confident. And if you're still sick with nerves, do what one friend of mine does: wear fabulous shoes!

Interviews

If public speaking is one of the things that people fear the most, interviews must come a close second. And for similar reasons, all related to negative thinking. Thinking such as 'I'll dry up', 'I'll talk too quickly' or 'I won't know the answers to the questions.'

there is no substitute for preparation

But, just like public speaking, for an interview there is no substitute for preparation.

First, do your research. Research the company/employer's services, markets, competitors and trends. This information will prove useful for two reasons. First, when you're asked the inevitable questions about why you want to work for the company. Second, you may be competing against an internal applicant, so knowledge of the company will limit their advantage over you.

On the day before the interview, read your application form again. Many of the questions you will be asked will arise from what you have written in your application. You need to remember what you have written and think about what questions might come out of what you have said.

For a positive first impression, when you meet the interviewer, shake their hand firmly, look them in the eyes and say 'hello'. These first few seconds are so important in creating a favourable impression. Keeping your body language natural and positive (smiles, head nods, good posture) will help you to project a confident image.

brilliant tip

Practise shaking hands. A limp handshake signals low confidence and self-esteem. On the other hand, an excessively strong handshake will communicate an aggressive approach.

Ask a friend, someone who is confident and assertive, to give you honest feedback. Keep practising until you get it right.

Listen. Don't let your nerves, expectations or prejudices influence how well you listen. Listen for feelings and values as well as for facts. Avoid formulating a response in your head, or interrupting while the interviewer is still speaking.

Clarify. If you are uncertain what the interviewer is asking you, repeat their question back to them. For example, the interviewer might ask, 'How do you think you would cope with the challenges?' To which you might ask, 'Are you asking me how I would cope with the day-to-day challenges of the job, or with the challenges of working for a company that is in the middle of restructuring?' This approach will show that you will have the confidence to seek clarification in the job if and when necessary.

Answer questions in a confident, firm voice. Don't mumble or rush or be too hesitant. Follow the question-answering rule: be specific, give examples and be positive. Answers should not be one word or one sentence, but equally should not go on too long.

For example, in answer to the question, 'What skills and strengths will you bring to the job?' you would need to think of actual experiences that illustrate the qualities you would bring to the job. So you might answer, 'My main strengths for this job would be patience and a calm and flexible approach. For example, in my last job, whenever we were close to a deadline, I was able to remain calm, which helped me to think clearly and to suggest ways we could rethink our priorities in order to meet the deadline.'

brilliant tip

Positive communication

One of the interview questions that you can prepare an answer to is, 'Why do you want to leave your current job?' It may be that you are bored to tears by the work and hate your manager, but don't say that. You should never criticise your current employer or manager, otherwise you'll come across as a moaner, a whinger and a whiner. The interviewer will want to hear positive answers that, for example, reflect an interest in your career progression and development. So you might say that having been in your current role for some time, you now want some new challenges and extra responsibility.

If you don't know the answer to a question, say so. Never lie.

Before the end of the interview you will probably be asked, 'Are there any questions you would like to ask?' You may well have some questions, but it's also possible that you have been given all the information you need and everything has been covered. If so, say so.

Meetings and groups

Meetings *can* play a useful role in getting people together to make something happen. Too often, however, meetings provoke more dread than excitement. Although the responsibility for organising and running a meeting lies mostly with the person conducting it, as a member of the group, you *do* have a role to play. By taking the opportunity to practise using effective communication skills, you can make sure that something worthwhile is achieved.

- Be prepared. Before the meeting takes place, make yourself aware of the point of the meeting: the agenda, aims, expectations and the issues being addressed. Read the agenda.

- Go with a purpose: to say something and/or to learn something.
- If you're not sure if you'll have anything to contribute, listen. Listen and maybe you'll learn something interesting or useful.
- Take the opportunity to practise reading people's body language and other forms of non-verbal communication. Does their non-verbal communication match what they say?
- If you find it difficult to sit still and listen, take notes. Scribble down ideas, flashes of insight and questions.
- As always, if you have something to say – questions, ideas or information – be specific. Decide what exactly you want to say. Give examples to back up your points or ideas.
- If other people aren't being heard say, 'I'd like to hear what ... thinks.' Ask the quieter members for their responses, ideas and opinions.
- If you are unclear, paraphrase what you have heard and ask for clarification.
- Leave the meeting with a clear idea of what will happen next, what people have committed themselves to and what the intended outcome is.

How to write better emails

Emails are generally shorter than letters. They are usually written to request something or in response to a request or question. Informal English, abbreviations and absence of standard greetings are common in emails.

But, like letters, emails need to include a clear reason for writing: the main point, or points. They need to be polite and easy to understand.

Subject line

Use a descriptive subject line. Explain what the email is about in a few words. This introduces the content of your email. Be short but specific. Instead of writing 'Urgent', write 'Meeting at 10.30 about pay rise'. Avoid writing subject lines such as 'Hello' or 'Good news' or 'Message from Jane', as some email providers automatically delete these as spam.

Use a new subject line each time you start a new topic, rather than hitting 'reply' to a past email on a different subject.

Greetings

Of course, you can copy the sender and use the same greeting that they have; it's fine to echo the greeting used by the person with whom you're emailing. But if you are the one to write first, here are some possible greetings.

'Hi' or 'Hi + first name' are informal, and you can use this for friends and colleagues. You can also use 'Hello' or 'Hello + first name'.

More formally, whether to use the recipient's first name rather than surname will depend on who you are emailing. In business and work situations, if you have communicated with the receiver previously and they are at a similar level to you, then the use of their first name would be appropriate. If the receiver is more senior to you, or if you are in doubt, it would be best (particularly in the first communication) to use the person's surname together with a title – for example, 'Hello Mr Botha'. If you are writing to a group of people, you could write 'Hello all', or 'Hello everyone'.

Some people use 'Dear', then the recipient's name. That is appropriate for a business letter, but it's too quaint for a business email. It is also too intimate for a business email. If you do insist on using it, to avoid being overfamiliar, you must balance it by using 'Mrs', 'Ms' or 'Mr'. So, instead of 'Dear Louise', it should be 'Dear Mrs Campbell'.

Often, people who would never in a letter write 'Nina' or 'Jack' or 'Mr Murphy' with no introductory word first, do not hesitate to do so in an email. But, even in an email, it can appear quite abrupt to bark out someone's name like that!

Finally, having no greeting at all is usual if you are having an email conversation – that is, a succession of emails going back and forth between you and another person. Using email in this way is almost like texting. In these situations, you don't need to write any greeting or name.

Starting your email

Start with a clear indication of what the message is about. Here are a few of the most common reasons for writing an email, along with some sentences you can use to start.

1. Replying to a previous email or other contact.
   'Thanks for your phone call.'
   'Thanks for meeting with me yesterday.'
   'Thanks for the information.'

2. Making a request.
   'Can you tell me …?'
   'Would you be able to …?'

3. Giving brief updates.
   'Just to let you know …'
   'I'm emailing to update you on …'

4. Changing plans.
   'Sorry, but I won't be able to meet you next week.'

5. Referring to an attachment.
   'Thought you might find the attached file of interest.'

Your email should then go on to give the full details in the subsequent paragraph(s).

Try to keep to only one point in each email. Often, if you include too many points, the other person simply replies to what they think is the most important point and genuinely (or conveniently) forgets to respond to the other points. (Be aware of this when *you* are replying to emails; check that you have replied to every point.) If necessary, you can send more emails on different points. Simply say, 'I'm going to send you another email next, about …'

Layout

If you do have more than one reason for writing, give each reason, idea or subject its own paragraph. The extra space helps your reader to understand you have more than one reason for writing. Use plenty of space; don't make the reader have to work to read your email.

lists are a good way to split up information and make it easier for the reader to understand

Lists are a good way to split up information and make it easier for the reader to understand. For example:

When you come to the meeting, please bring:

- *examples of your work*
- *ideas for future projects*
- *examples of competitors' work (so we can see what we're up against!)*
- *information about any legal issues we might need to consider.*

Use bullet points in lists; they draw the reader's attention to the fact that each one is a separate point.

Be careful what you write!

Keep your emails informative and polite. Remember that the reader could forward (deliberately or not) your email to other people, so only write what you would be happy for other people to read. (No gossip, no personal comments, no offensive comments, no confidential information and no sarcasm.) If in doubt, phone or have a face-to-face conversation instead.

To be safe, assume that all your emails will be read by everyone in the country!

Angry emails

If you are angry or upset with the recipient of your intended email, start with a new blank email and do not fill in the 'to' or 'address line'. This way you avoid inadvertently sending the email, only to regret it when you've calmed down. Type up how you feel and why. Vent all your feelings. Do not send! This is especially important with email, where you can type up and send an angry email as fast as you can throw a shoe!

When you *do* respond for real, keep to the facts and make 'I' statements rather than accusative 'you' statements. ('*I* am upset by your comments' rather than '*You* have upset me.')

Try never to say 'never'. It's better to move to an unequivocal position gradually. More often than not, you can reach an acceptable compromise along the way.

If the issue is sensitive, or the email long, ask someone else to read it too. Ask them if they would be offended by such an email if it were directed at them.

Endings

Common endings are 'Best wishes', 'Regards' and, when relevant, 'Thank you'. 'Yours faithfully' and 'Yours sincerely' are usually too formal for an email.

Finally, ask yourself these questions: 'Is this clear?' 'Does my reader know what to do next?' 'Is this polite?'

Leave enough time for a response. Recognise that requests that may take only a few seconds to write and send may take much longer to fulfil. But if more than a few days have passed and you haven't received a response, forward your original email and politely ask if your recipient received it and has had time to consider what you wrote. It may be more effective to follow up by phone.

Receiving emails

When you receive a reply, *acknowledge it*. A simple 'Thank you' may be enough. In fact, when you receive any email, unless you specifically intend not to reply, ever, *always reply*, if only to say that you received it and you'll give the sender a full answer in a couple of days, when you've got the relevant information they need, or you have, for example, got through a particularly busy period at work.

brilliant communicators *always* acknowledge the emails that they receive

Brilliant communicators *always* acknowledge the emails that they receive. So, if you want to avoid being perceived as flaky, inefficient, unreliable and disorganised, *always reply*.

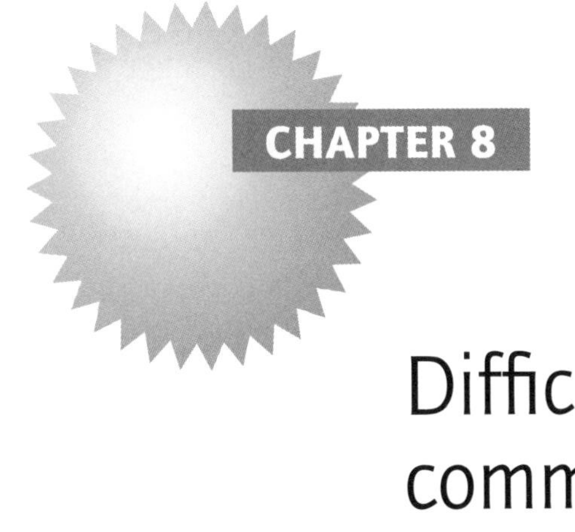

Difficult communication situations

When communication is going well, words, thoughts and feelings seem to flow effortlessly. But when things are going badly, the act of communicating can be as difficult as dragging a piano uphill. Here, we look at the communication skills and techniques you can use for communicating in difficult situations. Situations that include handling criticism and bad news are difficult for everyone concerned, but they can be handled sensitively. This chapter shows you how. It also considers a situation where the frustration is yours and yours alone: the other person is quite happily chatting away, oblivious to any discomfort on your part – but you want/need to leave. I show you how to shut someone up. Nicely.

First, here are some guidelines that apply to *all* difficult communication situations.

brilliant dos and don'ts

Do

- ✔ Listen. You may have missed some information that may have given you a different perspective on things. Be open to new ideas or facts that may change what you think and feel.
- ✔ Paraphrase and question to discover feelings. People rarely get their point across well in difficult situations. Paraphrasing helps you to be completely clear about

what they are saying. Reflecting someone's words back to them and asking if this is correct often highlights how ineffectively they explained something or that you didn't get the point at all. If you continue the exchange or act on something that you misinterpreted, things can get worse very quickly.

- ✔ Take time to recognise and acknowledge feelings and emotions. Too often, people try to hide their feelings in difficult situations. Bring them out in the open. Avoid doing this and it is more than likely that *emotions* will end up running the show.
- ✔ Empathise. Recognise how the situation might feel to the other person.
- ✔ Speak calmly, slowly and stay in control. Your speech, reactions and body language have an enormous impact on others, so you must be very careful with them to ensure you do more help than harm.
- ✔ Speak slowly. Be mindful of your speech and body language.
- ✔ Give evidence for your point of view. Not excuses or long-winded explanations
- ✔ Be honest. Be honest about what you do and don't know, what you think and feel and why. This can help give the other person a better understanding of where you are coming from.

Don't

- ✘ Claim to know exactly how the other person feels. You don't!
- ✘ Deny the other person's feelings or thoughts. ('I don't believe you feel like that' or 'No, that's not what you thought'.) This can make the other person defensive, and they will either lash out, or withdraw and say nothing more.

Giving and taking criticism: how to manage criticism

Who wants to listen to criticism? It usually hurts. In fact, the words you might use to describe having been criticised are pretty much the same as when you've been harmed physically; you may feel you've been 'laid into', 'blasted' or 'pounced on'.

If someone else's criticism pushes your buttons, you'll probably react emotionally, not rationally. It's hard not to react defensively, deny everything, blame someone else, counter-attack or sulk. You lose control and find yourself unable to listen objectively to the other person's remarks, calmly appraise them and respond accordingly.

Of course, your response to criticism is dependent on a variety of factors:

- *who* is criticising you
- *why* they are criticising you
- *when* they are being critical
- *where* they criticise you.

But whoever it is who's criticising you, whether it is fair or unfair criticism, there is a positive way to handle it. Using good communication skills, you can learn to handle criticism even if others aren't skilled in giving it! Once again, the emphasis is on active listening.

1. *Listen to what the other person is saying.* Really listen. Resist the urge to interrupt or apologise or do any of those other things that get in the way of really listening. Remember, the aim here is simply to *understand* what the other person is saying and what exactly they are criticising you about.

2. *Reflect.* Before you respond to the accusation, repeat back what you think the other person has said or is feeling.

For example, 'You're saying that I haven't done what you asked?' or 'Am I right, you're upset about the way I spoke to Joe?'

Strong emotions provoke strong opinions. Take time to recognise the feelings of the critic and you set the space up for a better understanding between you. Avoid doing this and it is more than likely that emotions will end up running the show.

3. *Ask questions.* If you're still not certain, ask questions to make sure you understand what has made the other person criticise you. For example, 'I just need to be clear; do you think I did it deliberately?'
4. *Respond.* If you're clear about what and why the other person is criticising you, now is the time to decide how you feel, and what you're going to do. You might want to refute what the other person has said, agree or negotiate. Whatever it is, say it. For example, 'I know that you're angry and you think I've been inconsiderate, but I do not agree because ...' Give evidence (not excuses) for your point of view.
5. *Repeat the process.* Listen to the other person's response and continue acknowledging and responding.
6. *Agree to disagree.* If you can resolve the situation, all well and good. But, if not, learn when to draw a line and agree to disagree.

brilliant example

Giving criticism

Adele: 'John, this work has *not* been done properly, it's *not* what I asked for. I'm *not* happy. You didn't do what I said. You're *hopeless* and now I'm going to have to spend time tomorrow getting it right'.

John: 'I'm sorry you're unhappy with what I've done. Are you saying that you are going to have to work on it because you think I'm not capable of putting it right?' (Notice that, at this point, John is neither agreeing nor disagreeing. He is simply reflecting and asking questions to clarify and establish understanding.)

Adele: 'Unhappy? I'm furious! YES! Yes, for God's sake! If you can't get it right the first time, you're obviously not up to it.'

John: 'OK. (John is speaking quietly and calmly.) How about we go through it together when you're not so angry. Then you could explain what exactly needs doing, and I can tell you if I feel confident about trying again. I'm new to all this, so I can see I probably didn't get it right. Before I go back to the task, I'll email you the details to make sure we're both clear and agree what needs to be done.'

Adele: (Calming down.) 'Hmm, OK. We'll need to spend an hour straight after lunch.'

Although criticism may come out more forcibly than the person intends, the message is still important. Reflective listening helps establish what the message is and how upset or angry the other person feels.

Reflective listening also helps to diminish the other person's attack. If the criticism is reflected back, the critic is more likely to feel understood and less likely to feel attacking and defensive.

> if the criticism is reflected back, the critic is more likely to feel understood and less likely to feel attacking and defensive

Giving criticism

Think back to the last time you criticised someone. Did you just want to let rip and get it off your chest so that you could feel better? Or maybe you were trying to be more constructive than that; you wanted to see a change in the other person's behaviour?

Constructive criticism can also be a form of feedback in that it provides information, opinion or analysis about something the person did which will inform positively their future performance or behaviour.

Whether you put it in writing or speak to the other person, there are several steps to consider. But, before you say anything, you must first decide just what the other person has done that's a problem for you, and then what change you want to see.

If you *don't* do this, you risk being ignored or starting a full-on confrontation!

Follow these steps.

1. *Say it!* Be specific. Say just what the other person did that has upset you, irritated you or made you angry. Then be specific about the change you want.

   It may be obvious to you what the problem is, but don't assume it's clear to the other person. Consider, for example, the time, energy and misunderstanding caused by this vague criticism: 'Emily, you must work on writing better emails. They need to be tightened up, OK?'

   A better approach might be: 'Emily, you made some interesting points in this email, but they were hard to find because of the lack of structure. Next time, could you bullet point each key issue and make it clear what your questions are?'

2. *Listen.* Listen to the other person's response. Do not interrupt.

3. *Acknowledge.* Respond to what the other person has said by repeating it. For example, 'OK, Emily, I know you

feel that all the information is there and I understand you don't think you've got the time to think about the exact wording. However, Bill is getting increasingly annoyed that he has to keep phoning you to clarify what you mean. And this takes up both your time and his. So please aim to get it right the first time. If you need help, let me know.'

4 *Listen to the other person's response.* Acknowledge it and either stick to what you want, or be prepared to negotiate or compromise.

Here are some other things to bear in mind.

- *Consider the time and place.* Do you have to say something immediately – or can it wait until the other person is more likely to listen to you?
- *Tell the other person how you feel.* Are you jealous, angry or upset? Don't be afraid to tell the other person how you feel. 'I was upset/embarrassed/furious when …'
- *Do not accuse.* Do not tell the other person, '*You* do this and *you* do that.' Instead, use 'I' statements. 'You' messages label the other person in a negative way. For example, instead of, 'You need to make it clear what you want' say, 'I would find it easier if you were clearer about what you want.'
- *Criticise one thing at a time.* If you have more than one issue, focus on one at a time and deal with the most important area first. Don't allow yourself to be dragged into other issues by the other person either.
- *Decide what your next step will be.* If the other person doesn't do what you want, what will you do? This doesn't mean issuing threats or punishments. It means deciding what your next step will be or what the *consequences* will be if they refuse to cooperate.

brilliant example

Gail: 'Liam, these factsheets are incomplete. Can you tell me anything about this?' (Focusing on the specific problem.)

Liam: 'Oh my God. I can never do anything right. I did what you asked. I've got a lot on at the moment. Really, sometimes I feel like giving up. Anyway, Tanya never even got hers in on time and you're not having a go at her, are you?'

Gail: 'I'm sorry you feel you can never do anything right and you are feeling overwhelmed. We need to go through the factsheets together and see if we can get them finished by the end of the afternoon.' (Acknowledging what John said but staying focused.)

Liam: 'Well, as far as I'm concerned, they *are* finished and I don't have any more time.'

Gail: 'I can see you're pushed for time, so that's why I'm suggesting we work together on this. What time would be best for you?' (Standing firm but also negotiating.)

Liam: 'Well, the only spare time I'll have is around 2 o'clock. But it had better not take long.'

Gail: 'Good. 'Thank you. We'll just keep working on it until we get it right.'

Giving bad news and negative feedback

Breaking bad news can be a real challenge; it can be difficult and distressing for both the giver and receiver of bad news. There are, however, ways to do it sensitively. What matters most is how well you listen and how you respond to the other person's reaction.

But first, if you have bad news to deliver, know that you must lay it out plainly. It's difficult if you're in a situation where you have to convey that sort of information, but the other person needs to be clear about the situation.

- Give reasons. Be prepared to explain why something happened and what caused it to happen. Avoid lengthy explanations.
- Anticipate questions and be prepared with concise and credible answers. If you don't have the answers, say so. If you know where the other person can get further information that might help to answer their questions, say so. If a question is complicated, rephrase it to simplify it, but without changing the meaning. If it's angry, recast it in neutral language. Keep control of your own emotions and answer questions with respect and sensitivity.

brilliant example

'Well Ed, I'm afraid it's not good news. Because you didn't get high enough grades, you won't be able to go to the university you chose this year.' (Pause.) 'But you will be able to resit your exams and reapply next year or see if the other universities you chose will give you a place.'

'I've got some bad news. I'm afraid your lovely dog Rex didn't make it after the operation. He died at 3 am. He was old and just not strong enough. But Rex wasn't by himself; the nurse was with him. I'm sorry.'

> the best way to give bad news is to say something positive before and/or after the bad news

- You may have been told that the best way to give bad news is to say something positive before and/or after the bad news. This is not to imply that things don't seem so bad, and certainly not to trivialise the bad news. The reason to include something positive is so that the other person has something positive to *focus* on. (As in the examples above, 'But you will be able to resit your exams and reapply next year or see if the other universities you chose will give you a place' and 'Rex wasn't by himself; the nurse was with him.'

- In some situations, you can soften the impact. One way to do this is to use the positive form of a sentence. For example, instead of saying 'We can't decide until tomorrow', write 'We will decide tomorrow.'
- Bad news is usually met with strong emotions. Acknowledge those emotions but try not to get emotional yourself (unless the bad news directly affects you, too). Supposing, for example, you had to dismiss an employee. You can't *not* dismiss them because they're crying. But you can acknowledge their distress: 'I'm sorry you're so upset. I can see this has come as a shock' or 'I can see this is painful for you.' An empathic response acknowledges not only someone's feelings but also the reasons for those feelings.
- Listen to the other person; let them talk. Validate their emotions, but primarily listen and acknowledge.
- Avoid saying 'I know just how you feel' or 'Try not to worry about it.' Although you might mean well, the other person might feel that you *don't* understand or you're attempting to bring things to a close.
- If it's relevant, state what, if anything, you can do to help, or ask 'Is there anything I can do?' Avoid over apologising. Instead, suggest possible actions. No one wants excuses – they want a solution, direction or tips on how to improve a situation, or ideas for the future. Focus on what can be done rather than what can't be done.

Should you always deliver bad news in person? It may seem easier to convey bad news via email or letter. Certainly, then you can plan exactly what to say and how to word it. You can also say what you want to without being interrupted. But you can't see how the other person feels and responds. And if that's the reason for putting it in writing, you've taken the coward's way out; you don't have to deal with the other person's response.

When you deliver difficult news in person, you can read the other person's body language and make appropriate adjustments to what you're saying. You can stop to clarify misunderstandings or to shed more light on issues of concern.

So, when you can, talk to the other person, face to face.

How to deliver bad news and negative feedback in writing

If you do have to give negative feedback or deliver bad news in a letter or email, be aware that just as the way you deliver bad news verbally can affect how it is received, the same is true when delivering bad news in writing.

Your introduction is important. It gives you an opportunity to set the context for the bad news or feedback. Context (the circumstances relevant to the issue) can make a difference to how bad news is grasped and understood. So, start with something positive. Describe what went well and why.

For example, here's a tutor giving written feedback to a student: 'Zoe, you have done really well with your essay, especially as you were unable to attend several classes due to illness.'

The aim is not to minimise the bad news so much as it is to place it in context and make it part of your message. You need to introduce it, and you need to leave the recipient with the bad news explained and in context. Only after you've laid the context and given the positive angle should you detail the negative news: 'Unfortunately, your essay is unfocused.'

Once the bad news has been delivered, it should be followed by an explanation, a solution or suggestion for improvement, or a course of action that will result in future prevention of the same problem, whichever is relevant. 'In future, decide what your main points are and keep checking that they always relate to the main question.'

If the bad news is the result of something you have done, offer a solution, such as some kind of corrective action or recompense.

> bad news is one thing. Dwelling on it makes it worse

Bad news is one thing. Dwelling on it makes it worse. If you put bad news in the proper context (as a small hitch on the journey), then you leave the reader feeling hopeful and more positive, while being completely honest about the unfortunate circumstances.

Conclude by showing that you care. It could be nothing more than an apology, but it should show genuine empathy.

What to do when someone gives you the silent treatment

Have you ever offended someone? Fallen out with a friend and now they're keeping their distance? Or maybe someone who's normally happy to speak to you is giving you the cold shoulder? Whether it's a family member, colleague or friend, how can you break their wall of silence without making things worse?

First, if you're not sure what's happened, ask yourself some questions. When did the silent treatment start? What happened that day, or in the days just before the behaviour changed?

Was it something you said, or something you did? Did you do something wrong or behave badly? Narrow it down to a few possibilities.

Whatever the reason, by cutting you off, the non-talker is trying to control the situation, protect themselves or punish you. How can you break through?

Fortunately, this is a situation where you can plan what you're going to say. It's easy to get nervous or defensive, or to come across the wrong way, so plan what you're going to say. Imagine you're alone with this person and say out loud what you want to say. Listen to the way you make your statement, and adjust your tone if need be. If the other person feels that you did something wrong, you want to make sure your tone doesn't indicate that you think they are being oversensitive; a snide or patronising tone will only make things worse between you. Positive body language is vital; folded arms, a tense posture and avoiding eye contact are not going to demonstrate an honest, open approach.

If you don't know what you've done wrong, say so. 'I feel like there's a problem between us and that you might be upset with me.'

Find out what and how the other person is feeling. Do they feel hurt or frustrated? Maybe they simply feel let down. Try asking, 'How do you feel about what happened between us?'

Listen and acknowledge what they say about how they feel. Explain how you feel, but be careful not to find fault or lay blame.

Take responsibility and apologise. This doesn't mean taking all the blame or suggesting that it's completely your fault. Identify what you do and do not feel responsible for. Admit what, if anything, you could have done differently and say that you're sorry it happened. Make it clear that you understand what you are apologising for.

For example, 'I was angry and shouted – I'm sorry I upset you by losing my temper over it' or 'I didn't do what you asked me to do – I'm sorry I let you down.'

Explain what you can do to put things right. You could, for example, say, 'I know that when I get angry and shout, you don't get a chance to have your say. In future, if I feel like I'm losing control, I'll leave the room until I feel calmer' or 'I know I let you down. Can I make up for it in some way?'

Try only once. If, after you have attempted to understand what the problem is and apologised, but you still can't get through, know that you have done your best.

Now, it is up to the other person to step up and begin communicating with you.

You need to decide if you are willing to leave the door open for the other person to come through once they feel like talking again. 'It really hurts that you're shutting me out, and I wish you would talk to me so we can move forward. I'm finding it difficult to deal with, so I'll have to stop waiting and just assume that you don't want to be friends any more. I don't want to do that, which is why I'm telling you now.'

Of course, it might be easier, or even the only option, to put it in writing. Again, take responsibility for what you could have done differently. Remember not to find fault or lay blame, but do acknowledge how you think the other person is feeling. Explain how you feel and what you are willing to do to put things right.

How to shut someone up. Nicely

Have you ever hesitated to begin a conversation with someone because you feared you'd never be able to end it? Do you often find yourself stuck with someone who talks about topics in needless detail? Maybe they always bring the conversation back to themselves? Perhaps someone you know repeats the same stories and anecdotes even when you say you've heard them before?

How can you take back the reins of the conversation without appearing rude? There *are* a number of ways to do it nicely. All of them require you to listen closely. By listening closely, you will find something you can pick up on and use to take control.

Make contact. Take a deep breath and make eye contact. Say their name. If it's appropriate, briefly touch their arm. Stand up if you were sitting down. Then lead the topic to a close in one of the following ways.

Summarise. Rather than switch off, use your listening skills to summarise something they have said. For example, 'So, Phil, trekking the Great Wall of China was obviously quite a challenge. I must do something like that myself one day.'

Give warning. 'Phil, I'd like to hear more, but I also need to talk to James before he leaves.' When you just have to get away, let the person know your time is running out.

when you just have to get away, let the person know your time is running out

Be nice. 'It was good to hear about your trip, Phil – you've inspired me. Good to talk with you.' You'll feel OK about ending the conversation and the windbag will be happier to let you go if you say something positive.

Explain your next move. 'Phil, I need to go to the loo/call my babysitter/catch James before he goes.' Just make sure you do what you say. Don't let the person see you were just making an excuse.

Widen the circle. 'Phil, come with me, I want to introduce you to James/go to the bar/get some food.' Whether the other person agrees to join you or not, this tactic makes them feel included, even if you then give them the slip. If you are in a group, try directing questions to someone else. Say: 'What do you think of this, James?' Or, 'I really want to hear what Olivia says about this.'

Of course, this all takes effort on your part, but these tactics will put you in control of the conversation, and taking control is better than being bored senseless!

Conclusion

The aim of this book is to help you learn the principles of communication; to communicate effectively and positively in a range of situations.

Throughout, I've also emphasised the *benefits* of developing and improving your communication skills.

I hope you will have recognised that the ability to communicate clearly can help you connect with people, build friendships and develop relationships at work and at home and in a range of social situations.

Having read this book, you will have discovered the principles of brilliant communication. I've included tips and techniques, practical pointers and examples to help you communicate easily and effectively in a wide range of situations.

A word of warning though – you won't achieve brilliant communication skills just by reading about them. If you want to improve your communication skills, you'll need to practise! Nothing will happen overnight – like all skills, communication skills require time and effort to develop.

Finding out your communication style and those of other people is a good place to start when you want to improve your communication skills. Take on small manageable challenges to develop your skills, such as talking to the checkout operator at the supermarket, or making a point of talking to someone you don't usually chat to at work. What's the worst that can happen?

You'll make mistakes; learn from them. Make sure you do learn from them though: don't over analyse a situation, but simply reflect on what went wrong and how you could do better next time.

Come back to the book every now and again to remind yourself of some of its ideas, tips and techniques. The more you practise, the more confidence and empathy you'll have and the better your communication skills will be.

Good luck!

Further reading

Mehrabian, A. (1981) *Silent Messages: Implicit communication of emotions and attitudes*. Belmont, CA: Wadsworth.

Remland, M.S. and Jones, T.S. (2005) 'Interpersonal distance, body orientation, and touch: The effect of culture, gender and age', *Journal of Social Psychology*, 135, 281–97.

Savitsky, K., Keysar, B., Epley, N., Carter, T. and Swanson, A. (2011) 'The closeness–communication bias: Increased egocentrism among friends versus strangers', *Journal of Experimental Social Psychology*, 47(1), 269.

Ting-Toomey, S. (1999) *Communicating Across Cultures*. New York: Guilford Press.

Index